Nanotechnology for Bioengineers

Synthesis Lectures on Biomedical Engineering

Editor
John D. Enderle, *University of Connecticut*

Lectures in Biomedical Engineering will be comprised of 75- to 150-page publications on advanced and state-of-the-art topics that span the field of biomedical engineering, from the atom and molecule to large diagnostic equipment. Each lecture covers, for that topic, the fundamental principles in a unified manner, develops underlying concepts needed for sequential material, and progresses to more advanced topics. Computer software and multimedia, when appropriate and available, are included for simulation, computation, visualization and design. The authors selected to write the lectures are leading experts on the subject who have extensive background in theory, application and design.

The series is designed to meet the demands of the 21st century technology and the rapid advancements in the all-encompassing field of biomedical engineering that includes biochemical processes, biomaterials, biomechanics, bioinstrumentation, physiological modeling, biosignal processing, bioinformatics, biocomplexity, medical and molecular imaging, rehabilitation engineering, biomimetic nano-electrokinetics, biosensors, biotechnology, clinical engineering, biomedical devices, drug discovery and delivery systems, tissue engineering, proteomics, functional genomics, and molecular and cellular engineering.

Introduction to Statistics for Biomedical Engineers
Kristina M. Ropella
2007

Capstone Design Courses: Producing Industry-Ready Biomedical Engineers
Jay R. Goldberg
2007

BioNanotechnology
Elisabeth S. Papazoglou and Aravind Parthasarathy
2007

Bioinstrumentation
John D. Enderle
2006

Fundamentals of Respiratory Sounds and Analysis
Zahra Moussavi
2006

Advanced Probability Theory for Biomedical Engineers
John D. Enderle, David C. Farden, and Daniel J. Krause
2006

Intermediate Probability Theory for Biomedical Engineers
John D. Enderle, David C. Farden, and Daniel J. Krause
2006

Basic Probability Theory for Biomedical Engineers
John D. Enderle, David C. Farden, and Daniel J. Krause
2006

Sensory Organ Replacement and Repair
Gerald E. Miller
2006

Artificial Organs
Gerald E. Miller
2006

Signal Processing of Random Physiological Signals
Charles S. Lessard
2006

Image and Signal Processing for Networked E-Health Applications
Ilias G. Maglogiannis, Kostas Karpouzis, and Manolis Wallace
2006

Nanotechnology for Bioengineers
Wujie Zhang

ISBN: 978-3-031-00540-4   paperback
ISBN: 978-3-031-01668-4   ebook
ISBN: 978-3-031-00047-8   hardcover

DOI 10.1007/978-3-031-01668-4

A Publication in the Springer series
SYNTHESIS LECTURES ON ADVANCES IN AUTOMOTIVE TECHNOLOGY
Lecture #60
Series Editor: John D. Enderle, *University of Connecticut*
Series ISSN
Print 1930-0328   Electronic 1930-0336

Nanotechnology for Bioengineers

Wujie Zhang
Milwaukee School of Engineering

SYNTHESIS LECTURES ON BIOMEDICAL ENGINEERING #60

ABSTRACT

Nanotechnology is an interdisciplinary field that is rapidly evolving and expanding. Significant advancements have been made in nanotechnology-related disciplines in the past few decades and continued growth and progression in the field are anticipated. Moreover, nanotechnology, omnipresent in innovation, has been applied to resolve critical challenges in nearly every field, especially those related to biological technologies and processes. This book, used as either a textbook for a short course or a reference book, provides state-of-the-art analysis of essential topics in nanotechnology for bioengineers studying and working in biotechnology, chemical/biochemical, pharmaceutical, biomedical, and other related fields. The book topics range from introduction to nanotechnology and nanofabrication to applications of nanotechnology in various biological fields. This book not only intends to introduce bioengineers to the amazing world of nanotechnology, but also inspires them to use nanotechnology to address some of the world's biggest challenges.

KEYWORDS

nanotechnology, nanofabrication, electrospinning, biosynthesis, nanomedicine, environment, bioengineering

Contents

Acknowledgments

The author gratefully acknowledges Dr. Anne-Marie Nickel for reviewing and Ms. Gina Mazzone for proofreading the manuscript drafts.

The author also acknowledges the financial support for this project form the Faculty Summer Development Grant of Milwaukee School of Engineering (MSOE).

Finally, the author appreciates the full support and love of his family members both in the United States and China, especially during the current pandemic, and pays homage to his mom, Aimei Nie (聂爱梅), who inspired him to be an educator.

Wujie Zhang
June 2020

CHAPTER 1

Introduction to Nanotechnology

1.1 NANOTECHNOLOGY DEFINITION

According to the National Nanotechnology Initiative, a U.S. Government research and development (R&D) initiative, Nanotechnology is science, engineering, and technology at the nanoscale, which ranges from about 1 to 100 nm. Nanotechnology is the study and application of extremely small things and is used across all other science fields, such as chemistry, biology, physics, materials science, and engineering [1]. There are many definitions of nanotechnology and the most commonly used one combines components of the above interpretations into the following definition, *the design, characterization, production, and application of structures, devices and systems by controlling shape and size at the nanoscale.* This definition is concise but comprehensive as it includes the design, characterization, and production aspects in addition to its applications. Moreover, devices and systems are used to replace nanomaterials which can be a vague and indiscriminate term.

Nano is a Greek prefix meaning draft. The comparison of different scales—macro, micro, and nano—is shown in Figure 1.1. It is generally agreed upon that nanoscale is from about 1 to 100 nm. In their 2004 report, the UK Royal Society and Royal Academy of Engineering suggested that the range of the nanoscale is from the size of atoms (about 0.2 nm) to 100 nm [2]. And it is widely accepted that 100 nm is the critical size for defining nanoscale. Within the nanoscale size range, materials show substantially different properties when compared with the same substances at larger scales [3].

1.2 HISTORICAL ASPECTS OF NANOTECHNOLOGY

Richard Feynman of the California Institute of Technology gave what is considered to be the first lecture on technology and engineering at the atomic scale, "There's plenty of room at the bottom," in 1959 at the annual meeting of the American Physical Society which then appeared as an article in the Caltech magazine, *Engineering and Science* [4]. This talk is considered to be the origin of nanotechnology. Although the concept of nanotechnology is relatively recent, human beings have been applying nanotechnological theory to their world for over a thousand years. Some early examples include the Lycurgus Cup (Figure 1.2) and stained-glass windows created during the 6th–15th century (Figure 1.3). Nanostructured materials were used in both

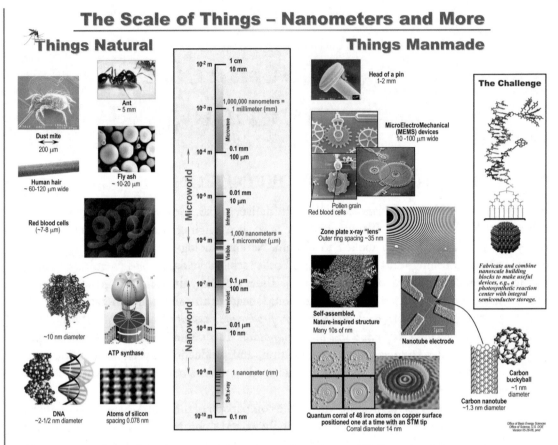

Figure 1.1: Scale of things (Source: Office of Basic Energy Sciences (BES), U.S. Department of Energy).

cases to create the effects desired by the artists. Colloidal gold and silver give the Lycurgus Cup unique optical properties. Gold chloride and other metal oxide nanoparticles are responsible for the creation of the vibrant colors seen in stained glass windows.

Other nanotechnology-related discoveries can be found throughout history. In 1857, Michael Faraday discovered colloidal gold (ruby-red). Semiconducting quantum dots in a glass matrix, buckminsterfullerene (C60, buckyball), and colloidal semiconductor nanocrystals (quantum dots) were discovered in the 1980s. In 1991, carbon nanotubes (CNT) were identified. CNTs show different structures (armchair, chiral, and zigzag; single wall vs. multi-wall) which offer their unique properties in terms of mechanical, electrical, optical, and others (Figure 1.4). Researchers at Rice University built a nanoscale car made of oligo (phenylene ethynylene) with alkynyl axles and four spherical C60 wheels in 2005 (Figure 1.5) [6].

Figure 1.2: The Lycurgus Cup (4th century AD). Left: in reflected light, and Right: lit from inside. Reproduced with permission, Copyright© The Trustees of the British Museum.

Figure 1.3: Glass window from Chartres Cathedral [5]. Reprinted from *Nanomaterials, Nanotechnologies and Design*, Michael F. Ashby, Paulo J. Ferreira, and Daniel L. Schodek, Chapter 2—An Evolutionary Perspective, pages 17–39. Copyright (2009), with permission from Elsevier.

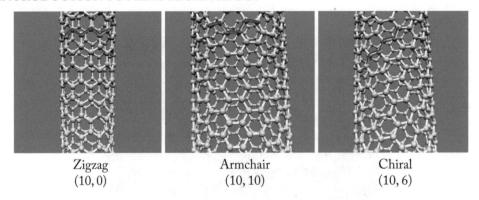

Zigzag
(10, 0)

Armchair
(10, 10)

Chiral
(10, 6)

Figure 1.4: Carbon nanotubes are classified as armchair, chiral, or zigzag based upon their structure. All the structures are generated using the Nanotube Modeler (Version 1.8.0 © 2005–2018 JCrystalSoft).

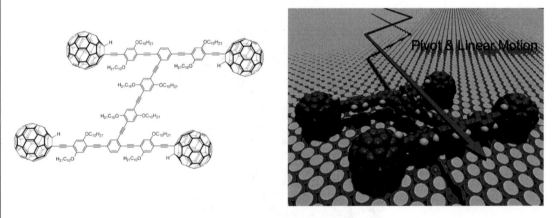

Figure 1.5: Structure (left) and methods (right) of motion of nanocar [7]. Reprinted with permission from Yasuhiro Shirai et al., *Nano Letters*, 5(11):2330–2334, 2005. Copyright (2005) American Chemical Society.

In order to make further advancements in any field, proper instrumentation is necessary to study different characteristics and the evolution of those instruments is just as important as the history of the field itself. When it comes to the instrumentation for nanocharacterization, Erwin Müller (Siemens Research Laboratory) invented the field emission microscope, allowing near-atomic-resolution images of materials in 1936. In 1981, two IBM researchers Gerd Binnig and Heinrich Rohrer (IBM) invented the scanning tunneling microscope (STM) and enabled individual atoms to be seen for the first time. The atomic force microscope, which allows for the viewing, measuring, and manipulation of materials down to fractions of a nanometer in size, was invented by Gerd Binning, Calvin Quate, and Christoph Gerber in 1986 [6]. From the atomic force microscope came the invention of the electron microscopes, such as SEM (scanning electron microscope) and TEM (transmission electron microscope), which used accelerated electrons as a source of illumination. Electron wavelengths can be up to 100,000 times shorter than that of visible light photons, and as the wavelength of the lamination source limits resolution, with the highly reduced wavelength, the resolution of a TEM is about 0.2 nm.

In the early 2000s, a clear and defined study of nanotechnology began. Different countries/unions started nanotechnology initiatives, including U.S. President Clinton, who launched the National Nanotechnology Initiative (NNI) in 2000. Nanotechnology education became more formalized and focused around the same time. The College of Nanoscience and Engineering at SUNY Albany was the first college-level education program; this program is now a part of SUNY Polytechnic Institute [8].

CHAPTER 2

Nanofabrication I: Electrospinning of Nanofibers

2.1 INTRODUCTION TO TOP-DOWN NANOFABRICATION

Nanofabrication is the manufacture of materials with nanometer dimensions. There are two general approaches to nanofabrication, top-down and bottom-up. Top-down nanofabrication is a subtractive process (from bulk material to nanoobjects with desired shape and size). Bottom-up nanofabrication is an additive process (from atoms/molecules to nanoscale objects, mostly chemical) [9]. Common nanofabrication approaches have been summarized in Table 2.1. The focus of this chapter is electrospinning (top-down approach) for the production of nanofibers.

Rolling/beating is probably the oldest top-down nanofabrication approach, although it is generally rejected as a viable option for the manufacture of nanomaterials at this time. The creation of gold sheets is one classic example of using rolling/beating. Gold leaf can be hammered down through a process called goldbeating into extremely thin unbroken sheets, these sheets have been known to reach nanoscale thickness [10]. The process of ball milling also utilizes mechanical energy to create nanoscale materials. Laser ablation uses lasers as an energy source, such as in the production of carbon nanotubes. When creating carbon nanotubes, a laser is applied to vaporize a graphite target with a catalyst and then condensed on a cool target downstream [11]. The anodization of $Al(Al \longrightarrow Al^{3+})$, forming anodic aluminum oxide (AAO), in certain strong acidic electrolytes can be used to produce nanosized array of pores. Processes for producing AAO was first patented by the Boeing Co. in 1974 [12]. AAO has been widely used as a template for deposition of the uniform arrays of nanowires, for example in biosensing [13]. Finally, lithography technology, such as photolithography and electron beam lithography, has been utilized for creating patterns with nanoscale size.

2.2 INTRODUCTION TO ELECTROSPINNING OF NANOFIBERS

Electrospinning, an electrohydrodynamic process, utilizes viscous solutions or melts to produce micro/nano-scale fibers. As shown in Figure 2.1, an electrospinning setup consists of three major components: a high voltage power supply, a syringe pump with a spinneret (mostly a metallic

Table 2.1: Common top-down and bottom-up nanofabrication approaches

Top-down	Bottom-up
Rolling/beating	Chemical vapor deposition
Ball milling	Sol-gel nanofabrication
Electrospinning	Supramolecular chemistry
Laser ablation	Self-assembly
Anodizing	Bio(Green)synthesis
Nanolithography	

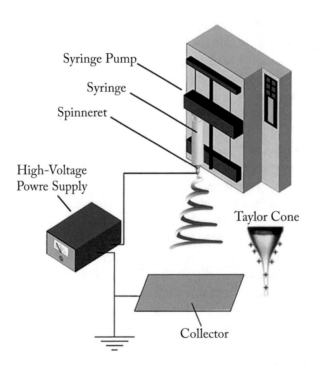

Figure 2.1: A typical electrospinning setup is shown in this figure and contains three major components: a high voltage power supply, a syringe pump with a spinneret, and a grounded collector [16]. Reprinted (adapted) with permission from Xue, J. et al., *Electrospun Nanofibers: New Concepts, Materials, and Applications, Accounts of Chemical Research*, 50(8):1976–1987, 2017. Copyright (2017) American Chemical Society.

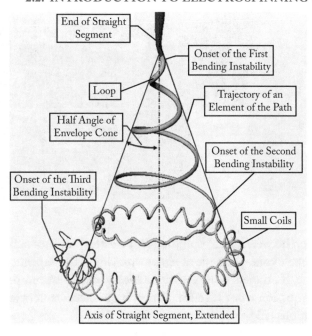

Figure 2.2: A diagram showing the path of an electrospun jet: end of the straight segment and development of bending instabilities [15]. Reprinted (adapted) with permission from Xue, J. et al., Electrospinning and Electrospun Nanofibers: Methods, Materials, and Applications, *Chemical Reviews*, 119(8):5298–5415, 2019. Copyright (2019) American Chemical Society.

blunt needle), and a grounded collector. Upon application of high voltage on viscous solutions or melts, a pendant droplet forms. When the electrostatic repulsion starts to overcome the surface tension of the fluid (by increasing the voltage), the pendant droplet will deform into a conical droplet, known as the Taylor cone, at the tip of the nozzle (connect to the positive end). As the electrostatic force overcomes the surface tension of the conical droplet (at a critical voltage value), a fine, charged jet is ejected from the tip of the needle. The interaction between the electric field and the surface tension of the fluid stretches the jet stream and makes it undergo a whipping motion (bending instability mainly) leading to the evaporation of the solvent and stream thinning (Figure 2.2). As the jet is stretched into finer diameters, it solidifies quickly, leading to the deposition of solid fiber(s) on the grounded collector [14–16].

Often there is confusion between the processes of electrospray and electrospinning, as the two processes share similar principles and setup. In general, electrospinning involves a higher viscosity polymer solution and voltage compared with electrospray [17]. During electrospray, the changes in parameters can result in the break of the jet into droplets rather than steam. The jet issuing from a Taylor cone experiences a range of competing instabilities, including the surface tension driven Rayleigh–Plateau instability, and the electrically driven axisymmetric conducting

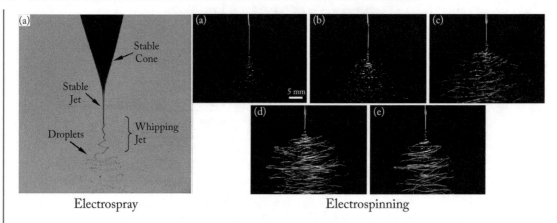

Figure 2.3: Comparison between electrospinning and electrospray production. Left: high-speed camera image of the stable cone-jet mode of electrospraying [20]. Reprinted from *Materials Today*, vol. 14, Jayasinghe, S., Biojets in *Regenerative Biology and Medicine*, pages 202–211. Copyright (2011), with permission from Elsevier. Right: high-speed camera jet images of different electrospinning conditions [21].

instability and whipping/bending instability. During electrospray, Rayleigh–Plateau instability dominates the process and the jet breaks up to form highly charged fine particles/beads. The viscoelasticity of the polymer solution competes with Rayleigh instability and is conventionally determined by the polymer concentration or molecular weight [18].

When it comes to the collector, aluminum foil can be used. However, specific collectors are required to align the nanofibers, such as drums and mandrel (Figure 2.4). Coaxial electrospinning has been developed and widely used to produce core-shell structured fibers. A coaxial nozzle is used as a spinneret. Bioactive molecules are included in the core when it comes to drug delivery. For example, it has been shown that the drug-loaded core-shell fibers had a better therapeutic effect on breast cancer than the use of the anticancer drug alone [19].

2.3 NOVEL PECTIN-BASED NANOFIBER PRODUCTION AND CHARACTERIZATION

Both synthetic and natural polymers have been used in electrospinning (summarized in Table 2.2) beside ceramics and composites. Natural polymers are generally preferred due to their biocompatibility for biological applications. However, the cost of natural polymers is higher than synthetic products. Certain properties, such as mechanical properties, of natural polymers, are not always comparable to those of synthetic polymers [23].

Among natural polymers, pectin has attracted more attention and has been applied in various fields including electrospun nanofibers, drug delivery, artificial red blood cell creation

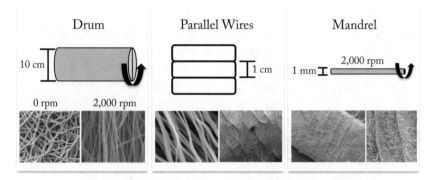

Figure 2.4: Different types of collectors with different electrospun nanofiber alignment. For a drum-type collector, the fiber alignment highly depends on the rotation speed [22]. Reprinted from *Materials Letters*, vol. 193, Manuel A. Alfaro De Prá et al., Effect of Collector Design on the Morphological Properties of Polycaprolactone Electrospun Fibers, pages 154–157. Copyright (2017), with permission from Elsevier.

Table 2.2: Synthetic and natural polymers used for electrospinning for biological applications

Synthetic	Natural
Polycaprolactone (PCL)	Collagen
Polylactic acid (PLA)	Fibrin
Polyglycolic acid (PGA)	Gelatin
Polyethylene Glycol (PEG)	Elastin
Poly(lactic-co-glycolic acid) (PLGA)	Fibrinogen
Poly-L-lactide-co-ε-caprolactose (PLCL)	Alginate
Polyethylene vinyl acetate (PEVA)	Chitosan
Polyethylene terephthalate (PET)	Pectin

as oxygen therapeutics, and bioprinting for tissue engineering [17, 24–30]. Pectin is a heteropolysaccharide found in the cell walls of all land plants. Most of the pectin products used are produced by either citrus fruit or apple peels. Since pectin is composed of galacturonic acid with carboxyl groups (Figure 2.5), it is polyanionic, which could interact with divalent cations (such as Ca^{2+}) and polycation (such as chitosan) to form a hydrogel or polyelectrolyte complex. Based on the degree of methyl esterification, pectin can be classified into low methyl (LM) and high methyl (HM) pectin. LM pectin contains a higher percentage of carboxyl groups compared with HM pectin [27, 30].

In general, pectin alone cannot be used for electrospun fibers like similar polysaccharides [31, 32]. Carrier polymers and surfactants are also needed. Poly(ethylene oxide) (PEO) and

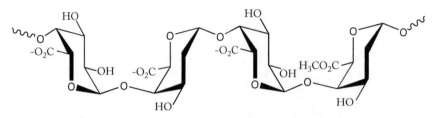

Figure 2.5: Chemical structure of pectin.

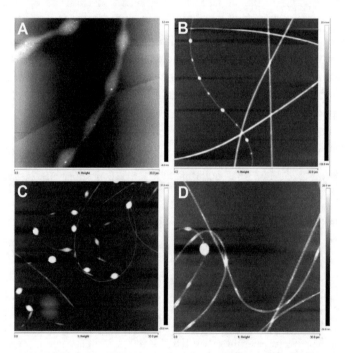

Figure 2.6: Atomic force microscopic images of pectin-PEO nanofibers with different pectin: PEO ratios. A: 70:30, B: 60:40, C: 50:50, and D: 40:60. Reproduced from reference [31].

poly(ethylene alcohol) (PVA) are commonly used as carrier polymers to increase the jet stability during electrospinning. Non-ionic surfactants, such as Triton X-100 and Pluronic F127, can be used to reduce the surface tension during electrospinning without influencing the solution conductivity. Carrier polymers and surfactants can be removed after electrospinning through selective washing if not desired [31, 32]. Polymer-to-carrier polymer ratio, voltage, and flow rate have been shown to be the significant factors influencing the electrospinning process. Figure 2.6 shows the influence of pectin, and how the PEO ratio influences the electrospinning process [31].

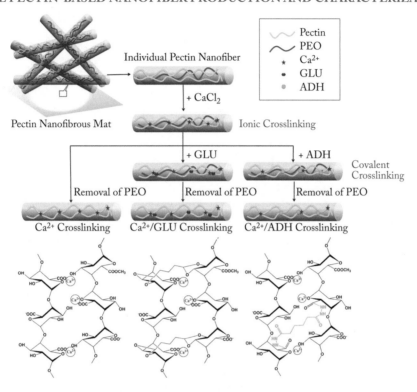

Figure 2.7: Schematic indicating the crosslinking and dual-crosslinking mechanisms of pectin-based nanofibers [33]. Reprinted from *Carbohydrate Polymers*, vol. 157, Cui, S. et. al., Effects of Pectin Structure and Crosslinking Method on the Properties of Crosslinked Pectin Nanofibers, pages 766–774. Copyright (2017), with permission from Elsevier.

Since pectin-based nanofiber is highly water-soluble, ionic or ionic/covalent dual crosslinking (Figure 2.7) is needed to improve its stability in aqueous solution. Both cationic ions and oligochitosan have been successfully used to crosslink the pectin-based nanofibers with oligochitosan shows lower-toxicity and provides a positively charged surface for promoting cell-attachment. $Ca^{2+}/$glutaraldehyde (GLU) dual-crosslinking improved the pectin-based nanofibers' mechanical strength; while $Ca^{2+}/$adipic acid dihydrazide (ADH) dual-crosslinking increased both mechanical strength and stability [33].

CHAPTER 3

Nanofabrication II: Biosynthesis of Silver and Gold Nanoparticles

3.1 INTRODUCTION TO SILVER AND GOLD NANOPARTICLES

As a consequence of their unique physicochemical properties, metal nanoparticles have been widely used in bioremediation, drug delivery, bioimaging, and other biological fields [34]. Among them, silver (Ag) and gold (Au) nanoparticles are the most studied. Silver nanoparticles show great catalytic activity and anti-microbial properties, they have been used in wound dressings and the treatment of cancer [35]. Although bulk silver is silver in color, nanoscale silver is golden (one good example of showing the significant material property change when reaching nanoscale). The color of Ag nanoparticles is size-dependent (Figure 3.1) [36].

Au nanoparticles can serve as a versatile platform for conjugation due to their large surface area-to-volume ratio and easy thiolate ligands grafting [37]. Additionally, Au nanoparticles can enhance optical signals, such as fluorescence and Raman scattering, for biosensing and bioimaging. Some of the plasmonic biosensing methods involving Au nanoparticles are localized surface plasmon resonance (LSPR) and surface-enhanced Raman spectroscopy (SERS) [38]. Interestingly, the optical properties of Au nanoparticles are both size and shape-dependent (Figure 3.2) [39, 40].

3.2 SYNTHESIS OF SILVER AND GOLD NANOPARTICLES

Colloidal metal nanoparticles are often synthesized by the reduction of metal salt or acid. For example, trisodium citrate ($Na_3C_6H_5O_7$) is a commonly used agent (both reducing and capping) for colloidal metal nanoparticle synthesis (Equation (3.1)). Other reducing agents include ascorbate, sodium borohydride, and poly (ethylene glycol)-block copolymers. The presence of surfactants and polymeric compounds, such as poly (vinyl alcohol), protects the particles from agglomeration and sedimentation (capping agent) [41]. For example, Y. Sun and Y. Xia synthesized Au and Ag nanoparticles using ethylene glycol served as both reductant and solvent and

Figure 3.1: Ag nanoparticle's color dependence on size (size increases from left to right) [36]. Reprinted by permission from Springer Nature Customer Service Center GmbH: Springer Proceedings in Physics (Bulavinets, T., Varyshchuk, V., Yaremchuk, I., Bobitski, Y., Design and synthesis of silver nanoparticles with different shapes under the influence of photon flows. In: Fesenko, O., Yatsenko, L. (Eds.), *Nanooptics, Nanophotonics, Nanostructures, and their Applications.*). Copyright (2018).

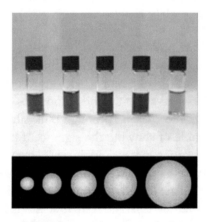

Figure 3.2: Au Nanoparticles of different sizes suspended in solution show different colors. (Courtesy of Aleksandar Kondinski.)

poly(vinyl pyrrolidone) (PVP) as a capping agent. And it was found that PVP played an important role in determining the shape and size of the nanoparticles [42]. Controlling the size and shape of metal nanoparticles is of great importance due to the strong correlation between these parameters and catalytic, antimicrobial, and other properties [42–44]. The chemical synthesis mechanism of Ag and Au nanoparticles is shown in Figure 3.3.

(See [45].)

$$Ag^+ + Na_3\,C_6\,H_5\,O_7 + H_2\,O$$
$$\longrightarrow Ag^\circ + C_6\,H_5\,O_7\,H_3 + Na^+ + H^+ + O_2 \ \uparrow \tag{3.1}$$

Biosynthesis of Au and Ag nanoparticles, such as the use of non-toxic chemicals, lower energy consumption, improved cost-effectiveness, and stable NP production. A large number of

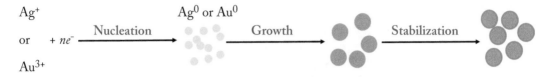

Figure 3.3: Chemical synthesis mechanism of colloidal Ag and Au nanoparticles.

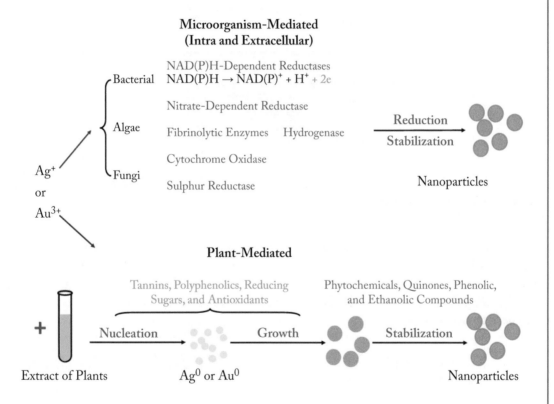

Figure 3.4: Biosynthesis routes of colloidal Ag and Au nanoparticles.

biological methods using microorganisms, including bacteria, fungi, and algae as well as plant extracts for Au and Ag NP production [46–48]. Biosynthesis routes of silver nanoparticles are summarized in Figure 3.4. When biosynthesis is plant-mediated, extracts of plants (different parts) are traditionally used. Different compounds within the extract serve as reducing and/or stabilizing/capping agents [46].

It is worthy of pointing out that certain organisms naturally produce nanoparticles. Take magnetic nanoparticles for example, magnetotactic bacteria (MTB) orient themselves and move

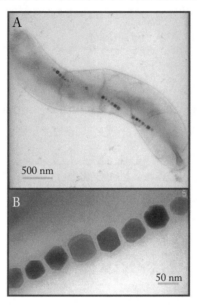

Figure 3.5: TEM images of magnetosomes in *Magnetospirillum Magneticum* AMB-1 cells. Figure reproduced from reference [51].

along magnetic field lines. They contain magnetosomes (iron-rich magnetic particles enclosed within lipid membranes) which are responsible for magnetotaxis in MTB (Figure 3.5) [49]. Furthermore, organisms have been used to produce other nanoscale substances, such as quantum dots (semiconducting nanocrystals). Biosynthesis of luminescent quantum dots in earthworms has been reported [50].

3.3   BIOSYNTHESIS OF SILVER NANOPARTICLES USING UPLAND CRESS

The leaf extracts of various plants contain the reducing agents necessary for silver nanoparticle formation. Green tea (Camellia Sinesis) [52], ginko tree (Ginko Biloba) [53], and magnolia tree (Magnolia Kobus) [53] and many other plants have been used for biosynthesis of nanoparticles [47]. Upland cress (Barbarea verna) belongs to the family of Brassicaceae (order Brassicales) and is widely available across the globe [54, 55]. Upland cress (Figure 3.6) exhibits excellent antioxidant capacity due to its high levels of antioxidants phytochemicals such as ascorbic acid, carotenoids, tocopherols and so on [56]. The advantages of biosynthesis of metal nanoparticles using upland cress are that it is easy to obtain and readily available and the low cost [47]. In addition, Triton X-114, a non-ionic surfactant, was used to purify the Ag nanoparticles by removing organic matter. The process for Ag nanoparticle synthesis using upland cress is shown

Figure 3.6: Picture of upland cress leaves. (Courtesy of Zachary Eckrose.)

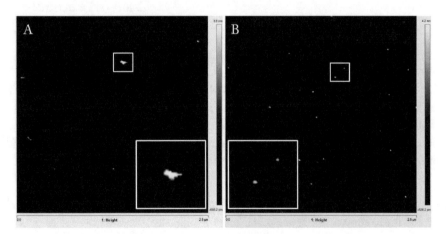

Figure 3.7: Ag nanoparticles production through chemical synthesis (left) and biosynthesis using upland cress (right).

in Figure 3.7. The differences between Ag nanoparticles with and without Triton X-114 treatment are shown in Figure 3.8. Lastly, the Ag nanoparticles synthesized using upland cress show antimicrobial properties against both Gram-positive and Gram-negative bacterial (Figure 3.9). When it comes to Ag nanoparticles' antimicrobial properties, different mechanisms have been proposed, causing physical changes in the bacterial envelope (cell wall and cell membrane), pro-

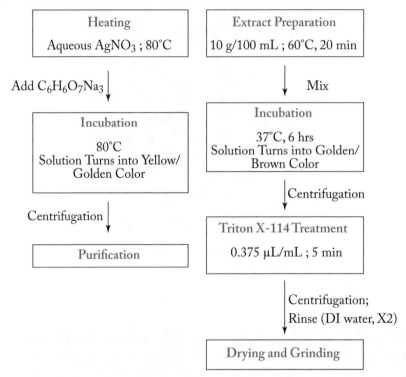

Figure 3.8: **AFM** images of silver nanoparticles without (A) and with (B) Triton X-114 treatment. Inserts are the enlargements of the selected regions [47]. Reprinted from *Micro and Nano Letters*, vol. 15, D. L. Johnson et al., Biosynthesis of silver nanoparticles using upland cress: Purification, characterization, and antimicrobial activity, pages 110–113. Copyright (2020), with permission from The Institution of Engineering and Technology.

ducing high levels of reactive oxygen species (ROS), and releasing toxic Ag^+ [47, 57]. The possible mechanisms are shown in Figure 3.10.

When it comes to nanoparticle size measurements, different techniques can be used. AFM, SEM, and AFM can show both morphology and size information directly. Dynamic Light Scattering (DLS) is also commonly used but mostly appropriate and accurate for more spherical particles as the determined size (hydrodynamic size) reflects the size of a perfect sphere that diffuses (Brownian motion) the same way [58]. Besides being used for confirming the crystal nature of the nanoparticles, X-ray diffraction analysis (XRD) can be used for determining the particle size. For example, the size of the silver nanoparticles (biosynthesize using upland cress) determined from the XRD data using the Debye–Scherrer equation (Appendix D) was in agreement with the results from AFM, DLS, and SEM analysis [47].

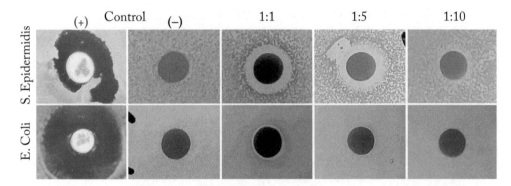

Figure 3.9: Antimicrobial results of synthesized silver nanoparticles. Different nanoparticle concentrations were tested: 1:1, 1:5, and 1:10 dilutions. The ampicillin disc and blank discs were used as positive and negative control, respectively [47]. Reprinted from *Micro and Nano Letters*, vol. 15, Johnson, D. et al., Biosynthesis of silver nanoparticles using upland cress: Purification, characterization, and antimicrobial activity, pages 110–113. Copyright (2020), with permission from The Institution of Engineering and Technology.

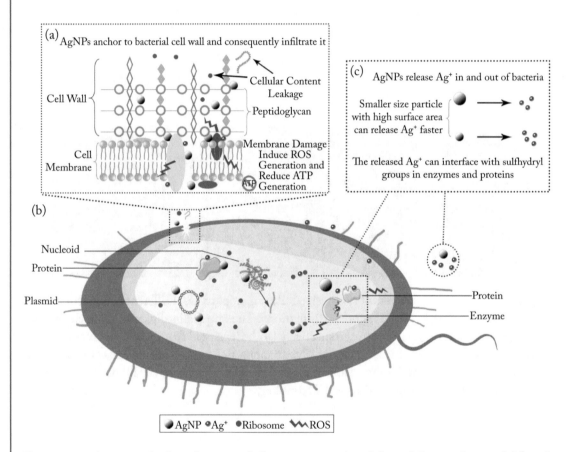

Figure 3.10: Antimicrobial mechanism of silver nanoparticles. Adapted from reference [57] with permission from Dove Medical Press Limited.

CHAPTER 4

Nanofabrication III: Self-Assembly

4.1 INTRODUCTION TO SELF-ASSEMBLY

Self-assembly (bottom-up nanofabrication) is a process in which particles or materials (modular construction units) at the nanoscale spontaneously arrange predefined components into ordered superstructures [59, 60]. There are two types of self-assembly, static, and dynamic. Static self-assembly is driven by energy minimization and doesn't dissipate energy. Dynamically driven processes require the dissipation of energy to sustain the structures or patterns [60]. The static and dynamic self-assembly processes can also be categorized into co-assembly, directed self-assembly, and hierarchical self-assembly. In general, co-assembly involves simultaneous self-assembly of different types of building blocks within the same system. Directed assembly requires external forces by design and hierarchical self-assembly is characterized by the organization of a single block over multiple length scales [60].

Molecular and nanoscale self-assembly is the form of self-assembly involving atoms, molecules, and crystal formation[61]. Different mechanisms of molecular self-assembly have been identified from basic molecular interactions to molecular-specific interactions [62]. These interactions are summarized in Table 4.1. Different stimulations, such as pH, temperature, ionic strength, and enzymes, have been applied/identified to promote molecular self-assembly [62].

4.2 EXAMPLES OF NUCLEIC ACID-BASED SELF-ASSEMBLY

Nucleic acid-based self-assembly is generally based on the base-pairing (hydrogen-bonds formation). Related progresses have significantly contributed to the formation of the emerging field of DNA/RNA nanotechnology. The invention of DNA origami is one of the milestones of the DNA nanotechnology filed. In this case, short DNA strands serving as staples to help the folding of long, single-stranded DNA into desired shapes [63]. Another design strategy is single-stranded tile assembly, in which the small building blocks for DNA nanostructure fabrication are called tiles while the single-stranded overhangs are termed sticky ends (allowing them to be linked together to form distinct topological and geometric features). Spatial factors, such as sticky ends orientation and flexibility of the double-helical domains, affect the assembly kinetics [64]. Mass production of DNA nanostructures is an emerging field, for example,

Table 4.1: Interactions involved in molecular self-assembly

Basic Molecular Interactions	Hydrogen bonds
	Electrostatic interactions
	Hydrophobic interactions
	π-π interactions
	van der Waals
Molecular-Specific Interactions	Nucleic acid base pairing
	Ligand-receptor biding

the production of single strands of DNA for DNA origami has been achieved in a liter-scale bioreactor[65]. Interestingly, self-assembly of DNA bricks, 13-nucleotide binding domains, into 3-D nanostructures of tens of thousands of unique components has been demonstrated recently [66].

4.2.1   DNA WALKER

DNA walkers have been designed, especially inspired by kinesin movement along a micro-tubule [68, 69]. A processive bipedal DNA nanomotor design is shown in Figure 4.2.

4.2.2   DNA HYDROGEL-BASED METAMATERIAL

Lee et. al. [70] reported a metamaterial with usual mechanical properties prepared using DNA as a building block. The materials exhibit solid-like properties in water but liquid-like properties out of water. The process involves a special polymerase Φ 29. This polymerase allows DNA chain elongation and displacement. Combined with a circular DNA template, both rolling circle amplification (RCA) and multi-primed chain amplification (MCA). Rolling circle amplifications enable long single-stranded DNA (ssDNA) formation, while multi-primed chain amplifications are responsible for waving (Figure 4.3).

4.2.3   DNA TRIPODS-BASED POLYHEDRAL

It has been recently demonstrated that various 3D polyhedrals can be self-assembled from DNA tripods [71]. The tripods of different inter-arm angles were designed and connectors (strands connecting the tripods) were used to assist the polyhedra assembly (Figure 4.4). Some of the applications of these polyhedral include the spatial arrangement of multiple enzymes [71].

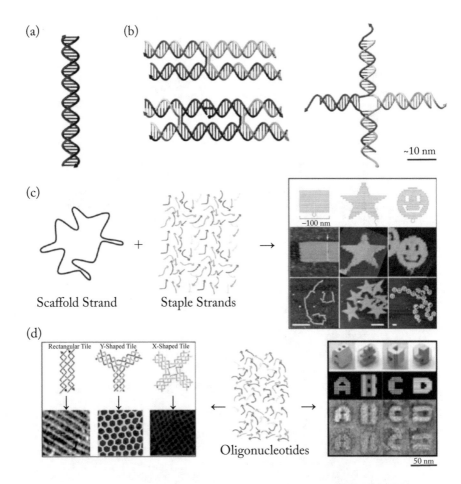

Figure 4.1: DNA-based structures. Structure of a DNA double helix (a). Single- and double-crossover motifs and Holiday Junction (b). Mechanism of DNA origami (c). Self-assembly of DNA tiles. Different types of DNA titles assembling into arrays (d) [67]. Reprinted (adapted) with permission from Mathur, D., Medintz, I. L., Analyzing DNA nanotechnology: A call to arms for the analytical chemistry community, *Analytical Chemistry*, 89(5):2646–2663, 2017. Copyright (2017) American Chemical Society.

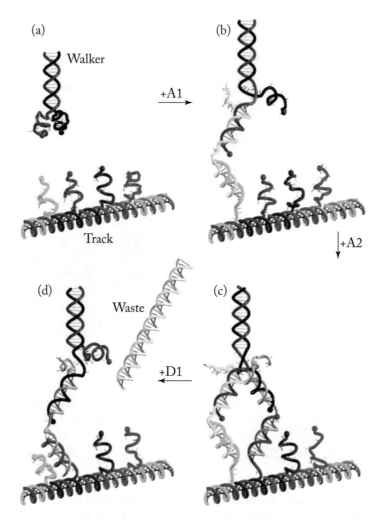

Figure 4.2: Schematic of walker locomotion. The (a) and (d) strands control the movement of the walker. The (a) strand specifically anchors the walker to a branch while the (d) strand is used to free a leg strand [68]. Reprinted (adapted) with permission from Shin, J.-S.; Pierce, N. A., A synthetic DNA walker for molecular transport, *Journal of the American Chemical Society*, 126(35):10834–10835, 2004. Copyright (2004) American Chemical Society.

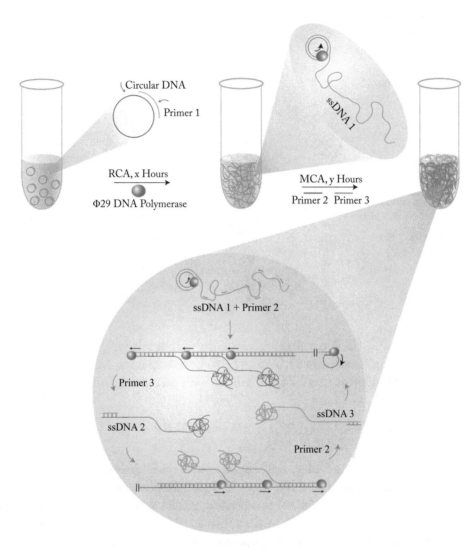

Figure 4.3: Process for synthesizing the DNA hydrogel. Schematic diagram of the stepwise approach for DNA hydrogel synthesis. First, long ssDNA strands (ssDNA 1) are synthesized based on a circular ssDNA template through RCA. The RCA and MCA processes were carried out as follows. Through MCA, two new types of ssDNA strands (ssDNA 2 and 3) are synthesized (ssDNA 2 and ssDNA 3 are complimentary while ssDNA 1 and ssDNA 3 have the same sequences) resulting in DNA hydrogel [70]. Reprinted by permission from Springer Nature Customer Service Center GmbH: Nature Nanotechnology (A mechanical metamaterial made from a DNA hydrogel, Lee, J. B. et al.). Copyright (2012).

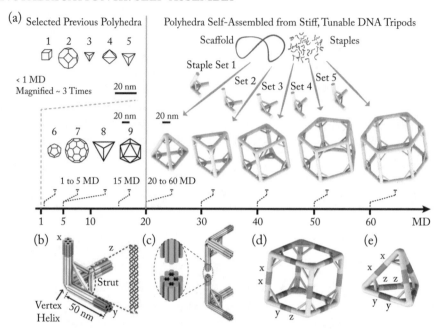

Figure 4.4: A: Polyhedra of different sizes and molecular weights. B: design of a tripod. C: Model showing the connection between two tripods. D and E: Schemes for assembling the cube and tetrahedron [71]. This figure is from [71]. Reprinted with permission from AAAS.

4.2.4   HELICAL SUPRAMOLECULAR POLYMERS VIA HYDROGEN-BONDED PAIRS STACKING

Bifunctional ureido-*s*-triazines with penta(ethylene oxide) side chains have been designed to achieve self-assembly into helical architectures in water. Inspired by the DNA double-helix structure formation mechanism, "linker" was used to overcome the interference of hydrogen-bond formation by solvent molecules to create a hydrophobic microenvironment. As shown in Figure 4.5, the stacking of aromatic units generates a hydrophobic microenvironment allowing for hydrogen bonding to occur within themselves rather than with water (solvent) [72].

4.3   EXAMPLES OF PEPTIDE/PROTEIN-BASED SELF-ASSEMBLY

Both peptides and proteins are made of amino acids. The essential difference between the two is "size." In general, 50–100 amino acid units are the cutoff. In 1993, first peptide nanotubes were synthesized by M. Reza Ghadiri and coworkers [73]. The building unit was an eight-residue cyclic peptide with the sequence: *cyclo*[− (D − Ala − Glu − D − Ala − Gln)$_2$ −]. Upon proto-

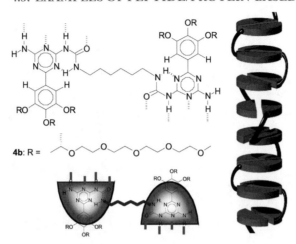

Figure 4.5: Propose mode of self-assembly of bifunctional unit 4b into helical polymers in water [72]. Reprinted (adapted) with permission from (Brunsveld, L., Vekemans, J. A. J. M., Hirschberg, J. H. K. K., Sijbesma, R. P., and Meijer, E. W. Hierarchical formation of helical supramolecular polymers via stacking of hydrogen-bonded pairs in water. *Proc. of the National Academy of Sciences*, 99:4977–4982, 2002). Copyright (2002) National Academy of Sciences, U.S.

nation, building units stack in an antiparallel fashion and produce a contiguous β-sheet structure through hydrogen bonding of the peptide backbone.

4.3.1   SELF-ASSEMBLY OF CYCLIC D, L-α-PEPTIDE

Cyclic peptides with an even number of alternating D- and L-α-amino acids show flat and ring-shaped structures. Cyclic peptides, under certain conditions, can stack to form hollow nanotubes (Figure 4.6). It has been proposed and proved that specifically-designed cyclic D,L-α-peptides may target and self-assemble in the bacterial membrane as antibacterial agents [74] and as modulators of plasma HDL (high-density lipoproteins) function [75].

4.3.2   ENGINEERED COILED-COIL NANOFIBERS FOR DRUG DELIVERY

Jin K. Montclare's group has engineered the coiled-coil domain of cartilage oligomeric matrix protein (COMPcc) by replacing cysteines with serines, fusing with elastin sequence motifs and so on [76, 77]. For example, a serine variant has been shown to be able to self-assemble into 10–15 nm nanofibers [78]. One application example of engineered coiled-coil nanofibers for drug delivery is encapsulation and release of a pan-RAR inverse agonist (BMS493) for osteoarthritis [76].

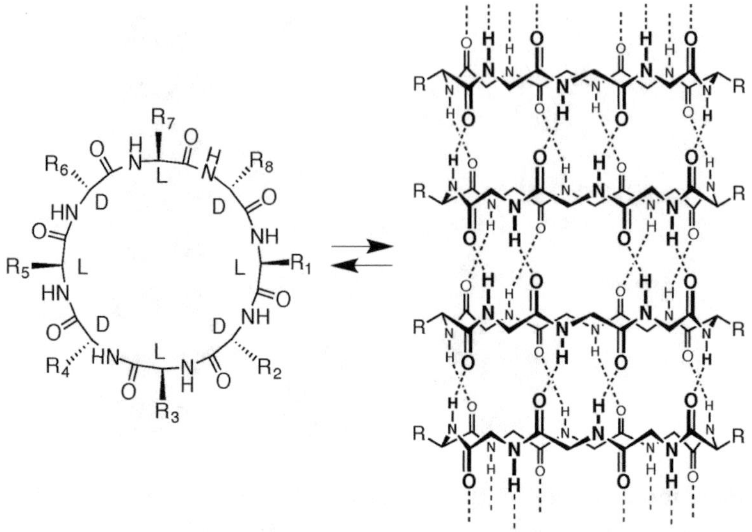

Figure 4.6: Structure of cyclic D,L-α-peptide (left) and self-assembled nanotube through hydrogen-bonding (right) [75]. Reprinted (adapted) with permission from Zhao, Y. et al., Self-assembling cyclic d,l-α-peptides as modulators of plasma HDL function. A supramolecular approach toward antiatherosclerotic agents, *ACS Central Science*, 3(6):639–646, 2017. Copyright (2017) American Chemical Society.

4.3.3 AMPHIPATHIC PEPTIDE SELF-ASSEMBLED NANOPARTICLES FOR GENE DELIVERY

McCarthy et. al. developed self-assembled peptide/DNA nanoparticles for non-viral gene delivery [79, 80]. The designed peptide (RALA) is composed of 30 amino acids. The hydrophobic leucine (L) regions interact with cell membranes, while hydrophilic arginine (R) regions facilitate nucleic acid binding. The peptide can condense DNA into nanoparticles. It is hypothesized that destabilization of the endosomal membrane could be explained by the increase in α-helicity content of the peptide in acidic pH [79].

4.4 EXAMPLES OF VIRUS-BASED SELF-ASSEMBLY

Self-assembly is widely used by biological systems, such as the formation of viral capsids. The RNA viruses' capsids can form spontaneously from the constituent coat proteins and RNA strands without ATP or other host-cell factors *in vitro*. Thus, self-assembly of capsids is considered to be driven by the minimization of the free energy [81, 82]. A recent study indicated that nucleation (with a concentration threshold) occurs on the RNA strands during self-assembly. One possible explanation is that growth is directed by the protein-RNA interactions [81]. Interestingly, the tobacco mosaic virus (TMV; rod-shaped) particle was the first macromolecular

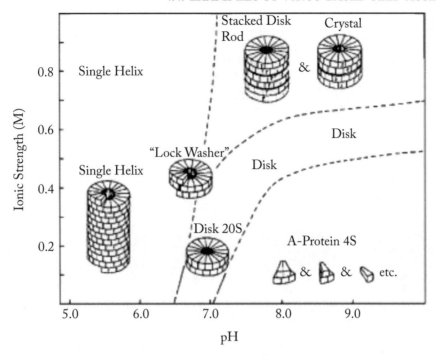

Figure 4.7: Different aggregated states of TMV coat protein in solutions of different pH and ionic strength [84]. Reprinted from *Biophysics Journal*, vol. 91, Kegel, W. K. and van der Schoot, P., Physical regulation of the self-assembly of tobacco mosaic virus coat protein, pages 1501–1512. Copyright (2006), with permission from Elsevier.

structure to be shown to self-assemble *in vitro* [83]. The coat protein of TMV can self-assemble into different aggregated states, such as disk-like assemblies and helices, in solutions depending on the pH, temperature, and ionic strength. Hydrophobic interactions, electrostatic interactions, and the formation of "Caspar" carboxylate pairs have been identified to be the three competing factors that regulate the transitions between these aggregated states [84]. Figure 4.7 shows the influences of pH and ionic strength on TMV coat protein aggregation. In addition, controlled self-assembly of the virus to form different dimensional materials have been extensively studied by tuning their surface nanotopography and polyvalent nature [85].

4.4.1   COWPEA MOSAIC VIRUS NANOPARTICLES (0-D)

Virus-like particle (VLP) has become an emerging research topic. VLP is defined as the spontaneous organization of viral coat proteins into the 3-D structure of a particular virus capsid [86]. VLPs possess inherent immunogenic properties and have been used for vaccination, such as Hepatitis B, influenza, and HIV [87], and cancer immunotherapy [86, 88–90]. Regard-

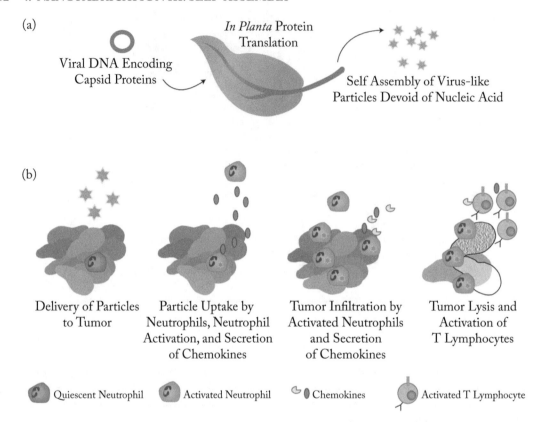

Figure 4.8: Production process and mechanism of application of cowpea mosaic nanoparticles for cancer immunotherapy [88]. Reprinted by permission from Springer Nature Customer Service Center GmbH: Nature Nanotechnology (A vaccine from plant virus proteins, Peruzzi, P. P. and Chiocca, E. A.). Copyright (2016).

ing cancer treatment, for example, cowpea mosic virus and *Macrobrachium rosenbergii* nodavirus (MrNVLP)-based VLP systems have been studied. As shown in Figure 4.8, empty CPMV (eCPMV) VLP system showed a significant reduction of established B16F10 lung melanoma and generated systemic antitumor immunity against B16F10 in the skin [86, 88]. VLP systems can be modified with targeting moiety (-es) and/or encapsulated with bioactive molecules. Qi-uXian Thong et al. modified MrNVLP with folic acid (FA; targeting cancer cells as many types of cancer cells express folic acid receptor (FR)) through Sulfo-NHS/EDC coupling reaction. Doxorubicin, a commonly used chemotherapy drug, molecules were encapsulated/retained in the VLP core by interacting with RNA (Figure 4.9).

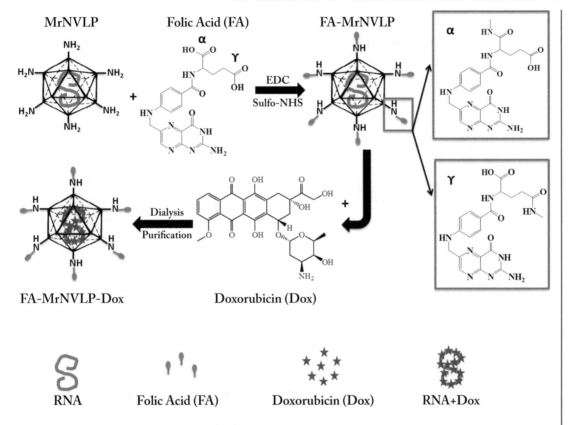

Figure 4.9: Schematic of MrNVLP-based drug delivery system. FA molecules are attached to the surface for targeting; while DOX molecules are encapsulated in the core through interaction with RNA. Adapted from reference [90].

4.4.2   ONE-DIMENSIONAL (1-D) POLYANILINE/TMV COMPOSITE NANOFIBERS

Self-assembly of TMV assisted by in-situ polymerization of polyaniline on the surface of TMV has been achieved to produce composite nanofibers and macroscopic bundled arrays [91, 92]. TMV can go through head-to-tail at suitable conditions and form a nematic liquid crystalline phase at high concentrations. As shown in Figure 4.10, polyaniline, as surface-coating, can prevent lateral association of TMVs resulting in long nanofibers at low TMV concentrations [91]. Aniline binds to TMV through electrostatic interactions and hydrogen bonding. The head-to-tail alignment is most likely contributed by the complementary hydrophobic interactions between the dipolar ends of the helical structure [93].

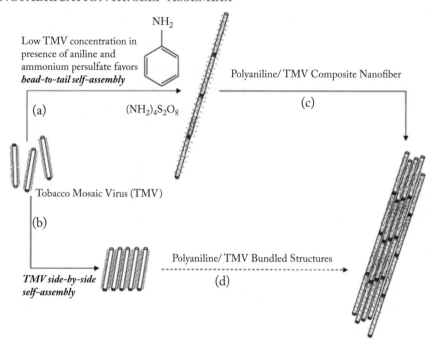

Figure 4.10: Polyaniline-assisted TMV nanofiber and bundled structure formation. A: head-to-tail self-assembly; B: TMV side-to-side assembly at high concentrations; C and D: possible mechanisms for polyaniline-coated TMV bundled structure formation [91]. Reprinted (adapted) with permission from Niu, Z. et al., Assembly of tobacco mosaic virus into fibrous and macroscopic bundled arrays mediated by surface aniline polymerization, *Langmuir*, 23(12):6719–6724, 2007. Copyright (2007) American Chemical Society.

4.4.3 TWO-DIMENSIONAL (2-D) BACTERIOPHAGE M13 THIN FILMS FOR ORIENTED CELL GROWTH

Rong et al. developed a thin film using M13 bacteriophages as basic building blocks. The film grafted with RGD (Arg-Gly-Asp) peptides shows the capability of guiding cell alignment and orienting the cell outgrowth (Figure 4.11) [94]. To form the film, M13 solution was first deposited on positively charged silane-coated glass slide (the coating is for strong binding of negatively charged M13 virus at neutral pH). With the constant slow dragging, the aligned virus thin film was formed as the solvent evaporates [94, 95].

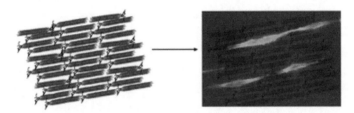

Figure 4.11: Schematic of M 13 bacteriophage thin film (left) and oriented cell growth on this film (right). Reproduced from Ref. [94] with permission from The Royal Society of Chemistry.

CHAPTER 5

Biomedical Applications of Nanotechnology

5.1 INTRODUCTION TO NANOMEDICINE

Applications of nanotechnology in healthcare triggered the formation of the new discipline—nanomedicine. Nanomedicine has sought to be a key enabling instrument for personalized but cost-effective medicine [96]. According to the U.S. National Institute of Health (NIH), nanomedicine refers to highly specific medical intervention at the molecular scale for curing diseases or repairing damaged tissues [97]. Nanomedicine has greatly impacted all fields of medicine from diagnostics to regenerative medicine.

5.2 NANOTHERANOSTICS

Nanotheranostics is the integration of diagnostic and therapeutic functions in one system using the benefits of nanotechnology [98]. A large number of nanocarriers have been designed for theranostics: these carriers are mostly nanoparticle or liposome/micelle-based [99]. Regarding the diagnosis, nanocarriers are generally incorporated, either attached to the surface or encapsulated, with fluorescent moieties, quantum dots (QDs), or radioisotopes for bioimaging. When it comes to drug/gene delivery, nanocarriers offer various advantages including easily tuned to achieve targeted delivery and controlled release, and improved solubility and stability [100]. Both diagnostic and therapeutic design aspects are summarized in Figure 5.1.

    There are two general approaches for nanocarriers: passive and active targeting. Passive targeting takes advantage of human physiology, e.g., (enhanced permeability and retention) EPR effect. EPR effect is the process that nanosized drugs leak preferentially into tumor tissue through permeable tumor vessels and are then retained in the tumor bed due to reduced lymphatic drainage [101]. However, the question has been raised as to whether the EPR effect is sufficient for cancer treatment [101] and EPR remains a controversial topic [102]. On the other hand, active targeting is typically based on ligand-receptor recognition [103, 104].

    Chemical modification with polyethylene glycol (PEG), i.e., PEGylation has been used to improve the pharmaceutical properties of both small molecule and biotherapeutic drugs. Particularly, PEGylation can help prolong the circulation time of the nanocarriers [105]. Intravenously administered nanomaterials are eliminated from the body mainly by renal and hepatobiliary elimination [106]. However, accelerated blood clearance (ABC) phenomenon, significantly de-

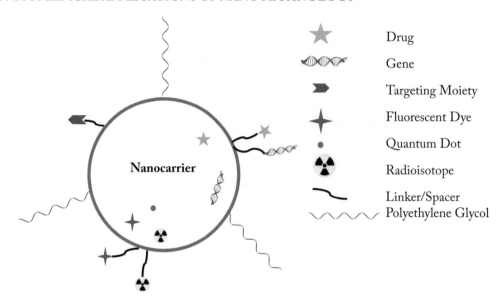

Figure 5.1: Nanocarrier design for theranostics.

creased circulation half-life, remains a challenge when PEGylatedd nanomaterials are repeatedly injected *in vivo* [107].

A good example is cancer treatment using ultrasound nanoteranostics [108]. Min et al. developed a doxorubicin-loaded calcium carbonate ($CaCO_3$) hybrid nanoparticles (DOX-CaCO3-MNPs). At tumoral conditions (acidic), these nanoparticles produce carbon dioxide bubbles showing great echo persistence, and release the drug at the same time (Figure 5.2) [109].

5.3   BIOSENSING AND BIOANALYSIS

Various promising nanomaterials (e.g., quantum dots (QDs), nanoparticles (NPs), and carbon nanotubes (CNTs)-based) have been widely used in biosensing and bioanalysis [110]. In particular, CNTs have been used for detecting various biological structures such as glucose, viruses, and so on [111]. CNTs have a large specific surface area which enables surface functionalization, exhibits unique intrinsic optical properties and unusual electronic properties, and could easily cross biological membranes (for potential *in vivo* applications) [111, 112]. Recently, a self-powdered photoelectrochemical glucose biosensor was developed by depositing supercapacitor CNT and Co_3O_4 onto the anatase TiO_2 coated electrodes (Figure 5.3) [113]. At 0 V, a linear response of the biosensor to 0–4 mM glucose was observed [113].

Another example regarding CNT-based biosensor is related to cancer detection. As shown in Figure 5.4, the biosensor for cancer detection is based on CNT and gold NPs. DNA detec-

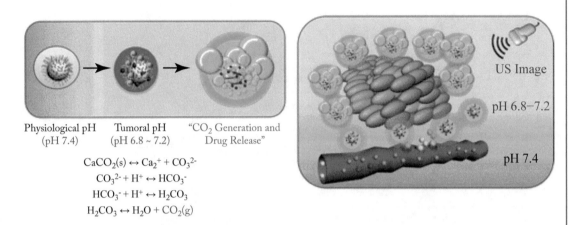

$$CaCO_2(s) \leftrightarrow Ca_2^+ + CO_3^{2-}$$
$$CO_3^{2-} + H^+ \leftrightarrow HCO_3^-$$
$$HCO_3^- + H^+ \leftrightarrow H_2CO_3$$
$$H_2CO_3 \leftrightarrow H_2O + CO_2(g)$$

Figure 5.2: A scheme showing the mechanism of CO_2 generation and drug release (left) and bubble generation for ultrasound imaging and drug release after the nanoparticle accumulation at tumor tissues (right) [109]. Reprinted (adapted) with permission from Min, K. H. et al., pH-Controlled Gas-Generating Mineralized Nanoparticles: A Theranostic Agent for Ultrasound Imaging and Therapy of Cancer, *ACS Nano*, 9(1):134–145, 2015. Copyright (2015) American Chemical Society.

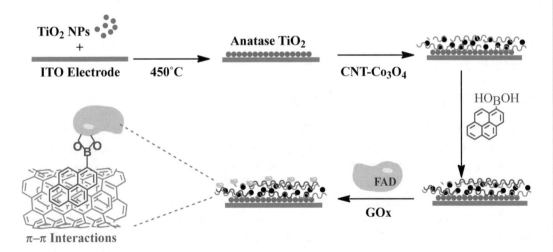

Figure 5.3: Biosensor fabrication process. Reprinted from *Biosensors and Bioelectronics*, vol. 119, Çakiroğlu, B. and Özacar, M., A self-powered photoelectrochemical glucose biosensor based on supercapacitor Co_3O_4–CNT hybrid on TiO_2, pages 34–41. Copyright (2018), with permission from Elsevier. GOx: glucose oxidase enzyme; FAD: flavin adenine dinucleotide.

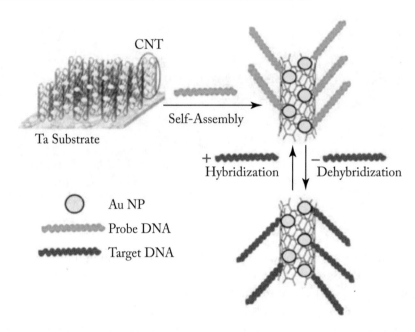

Figure 5.4: Schematic of the CNTs/NPs-based DNA biosensor for cancer detection. Adapted from reference [114].

tion targets mutation of the tumor suppressor gene TP 53. Gold NPs are used to improve the detection sensitivity by increasing the DNA probe density [114, 115].

5.4   TISSUE ENGINEERING

A classic definition of tissue engineering is "an interdisciplinary field which applies the principles of engineering and life sciences toward the development of biological substitutes that restore, maintain, or improve tissue function" [116]. Nanotechnology, including nanorobots [117], has greatly impacted all four aspects of tissue engineering: cells, scaffolds, bioactive molecules (e.g., growth factors), and bioreactors. In this section, the focuses are on the cells and scaffolds aspects.

Nanotechnologies have been applied to probe the nanoelasticity of cell microenvironments. It has been well studied that microenvironment affects cell behavior, such as differentiation [118]. For example, cell morphology and differentiation can be controlled by substrates with independently tunable elasticity and viscous dissipation [119]. AFM tips, especially various shapes, have been used and studied for nanomechanical properties measurements [120].

When it comes to scaffolds, scaffolds with nanoscale architectures are preferred as they exhibit large surface areas for bioactive molecules absorption and cell binding [121]. Electrospun nanofibers, self-assembled peptide-based nanostructures (e.g., ß-hairpin peptides [122]), and

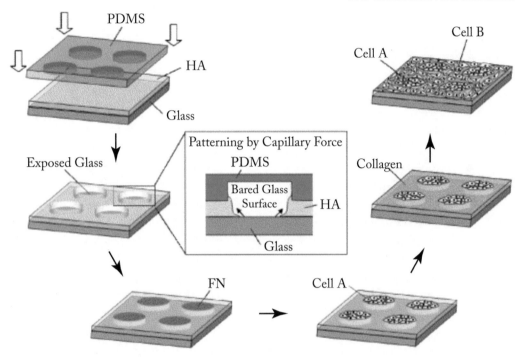

Figure 5.5: Scheme of the co-culture system fabrication process. Reprinted from *Biomaterials*, vol. 27, Fukuda, J., Khademhosseini, A., Yeh, J., Eng, G., Cheng, J., Farokhzad, O. C., Langer, R., Micropatterned cell co-cultures using layer-by-layer deposition of extracellular matrix components, pages 1479–1486. Copyright (2006), with permission from Elsevier.

self-assembled monolayers (e.g., alkanethiolate [123]) have been used for fabricating scaffolds. Interestingly, layer-by-layer (LBL) deposition has been used to control the surface properties of biological interfaces. For example, LBL deposition, using hyaluronic acid (HA), fibronectin (FN), and collagen, was studied to develop a cell co-culture system [124] (Figure 5.5). HA is cell-repellant *in vitro*; while FN and collagen promote cell adhesion [125].

Lastly, nanoglue is an emerging field for tissue engineering including surgery. Nanosilica solution has been demonstrated to be able to glue biological tissues together [126]. Moreover, fibrin glue, using fibrinogen, thrombin, and calcium, has been used to reconnect the severed optic nerve [127].

CHAPTER 6

Food Applications of Nanotechnology

6.1 INTRODUCTION TO FOOD APPLICATIONS OF NANOTECHNOLOGY

According to the United Nations Food and Agriculture Organization (FAO), the number of hungry people had been declining for decades, but this is no longer true. The number of undernourished people in the world has been on the rise since 2015, reaching 821.6 M in 2018 [128]. At the same time, food waste is estimated at between 30-40% of the food supply in the United States [129]. Food loss is also an issue. Food loss is defined as the decrease in the quantity or quality of food resulting from decisions and actions by food suppliers in the chain, excluding retailers, food service providers, and consumers [130]. Nanotechnology offers new opportunities to increase food production that may help those facing malnourishment and improve food production processes, food packaging, and food quality and safety [131, 132].

6.2 FOOD PRODUCTION

Currently, the applications of nanotechnology in food production focus on encapsulation of functional ingredients, structure building, and so on [133]. Nanoencapsulation of functional food ingredients can help increase their water solubility/dispersibility in foods and beverages, improve their bioavailability, mask undesired flavors/tastes, enhance shelf-life and compatibility, and control release rate or specific delivery environment [134]. For example, electrosprayed whey protein-based nanocapsules have been used for ß-carotene encapsulation [135]. Additionally, nanoencapsulation has been applied in antioxidants delivery in food systems, such as cinnamic aldehyde [136] and natural phenolic extracts [136].

Protein fibrils are promising building blocks for the preparation of macrostructures in food products and are usually used to modify the product properties [137]. Various protein sources have been studied such as whey protein [137], zein [138], and ß-Lacoglobulin [139] to produce nanofibrils. Interestingly, soy protein isolate fibrils showed a superior thermal protection effect on betalain than soy protein isolate [140].

6.3   FOOD PACKAGING

The main role of food packaging is the protection and maintenance of its integrity and quality of packed food [141]. Food packaging is classified as passive, active, intelligent/smart packaging [142, 143]. Passive packaging is a traditional packaging that involves the use of covering materials, characterized by some inherent insulating, protective, and ease-of-handling qualities [144]. Active packaging incorporates additives into packaging systems to maintain or extend food quality and shelf-life. This differs from intelligent packaging systems that monitor and communicate information regarding the quality of the packaged food during transport and storage using different indicators, such as ripeness [142, 143, 145].

Chitosan nanofiber and nano-formulated cinnamon oil have been incorporated into whey protein-based novel active packaging films [146]. Chitosan nanofiber was shown to be able to improve the mechanical strength and barrier properties and the nano-formulated cinnamon oil exhibited antibacterial activity [146]. Nano-ZnO was used to provide good antibacterial bioactivity [147]. Mulberry anthocyanin extract [147], thymol [148], tocopheryl polyethylene glycol 1000 succinate [149], and others have been studied in relation to antioxidant activities. Moreover, montmorillonite and derivatives have been used to improve the oxygen barrier properties—montmorillonite clays form a tortuous pathway for oxygen molecules to permeate through the packaging matrix [148]. Nano TiO_2 has been investigated for oxygen scavenging as it offers the ability to be photoinduced by UV radiation [150].

The concept of intelligent packaging development involving nanotechnology is shown in Figure 6.1. Indicators, sensors, radio frequency identification tags, and other systems are critical tools for intelligent packaging [150]. Take time-temperature indicators (TTIs), e.g., Zhang et al., designed a plasmonic TTI, which gives a shape red-to-green color change. As shown in Figure 6.2, the indicator is based on Au nanorods. The chemical reaction of epitaxial overgrowth of Ag shell on Au nanorods exhibits the chronochromic behavior. Microbial growth and the chemical self-evolution (chronochromic) are synchronized regardless of the T (Figure 6.2) [151].

6.4   FOOD SAFETY AND QUALITY

The Food Safety and Inspection Service (FSIS) of the U.S. Department of Agriculture, the U.S. Food and Drug Administration (FDA), and the Centers for Disease Control and Prevention (CDC) serve important roles in ensuring food safety in the United States [152]. FSIS is responsible for ensuring that the nation's commercial supply of meat, poultry, and processed egg products are safe, wholesome, and correctly labeled and packaged. The FDA assures that foods (except for these are regulated by FSIS) are safe, wholesome, sanitary, and properly labeled [152]. Very recently, the Agricultural Marketing Service (AMS) of USDA developed the List of Bio-engineered (BE) Foods to identify the crops or foods that are available in a bioengineered form throughout the world and for which regulated entities must maintain records [153]. It is worth

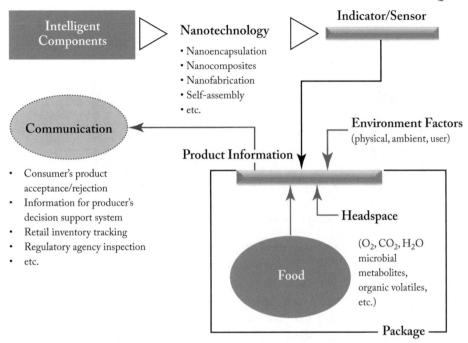

Figure 6.1: Scheme of the concept of intelligent packing development involving nanotechnology. Reprinted from *Trends in Food Science and Technology*, vol. 93, Kalpana, S., Priyadarshini, S. R., Maria Leena, M., Moses, J. A., Anandharamakrishnan, C., Intelligent packaging: Trends and applications in food systems, pages 145–157. Copyright (2019), with permission from Elsevier.

citing this information as applications of nanotechnology in bioengineering foods have been increasing. BE symbols are shown in Figure 6.3.

Nanosensors have been used in food safety and quality control for pathogenic bacteria, food-contaminating toxins, adulterants, vitamins, dyes, fertilizers, pesticides, and sensory detection [154]. Nanosensors offer serval advantageous properties including high accuracy and quick response [155]. Various types have been developed such as metal colorimetric nanoparticles detectors, carbon nanotube biosensors, nanocantilevers, array biosensors, and optical fiber nanosensors [155]. During this section, the focus is on aptamer nanosensors.

Aptamers are a class of folded nucleic acid strands capable of binding to different target molecules with high affinity and selectivity [156]. As a potent alternative of antibodies, aptamers are used to construct different types of nanomaterial-based, e.g., nanoparticles (NPs), quantum dots (QDs), carbon nanotubes (CNTs), and sensors (aptasensors) [157]. For example, a label-free fluorescent aptasensor for Alfatoxin B1 (AFB1) detection in food samples (Figure 6.4) was developed [158]. The system was based on quaternized tetraphenylethene salt (TPE-Z) and

Figure 6.2: A scheme showing the mechanism of the TTI: the chemical reaction (epitaxial overgrowth of Ag shell on Au nanorods) and microbial growth are kinetically synchronized (T-independent). The chemical reaction leads to color change from red to green (chronochromic) [151]. Reprinted (adapted) with permission from Zhang, C. et al., Time—Temperature Indicator for Perishable Products Based on Kinetically Programmable Ag Overgrowth on Au Nanorods, *ACS Nano*, 7(5):4561–4568, 2013. Copyright (2013) American Chemical Society.

Figure 6.3: Bioengineered symbol (left: https://www.ams.usda.gov/sites/default/files/media/Bioengineered.png) and derived from symbol (right: https://www.ams.usda.gov/sites/default/files/media/DerivedFrom.png).

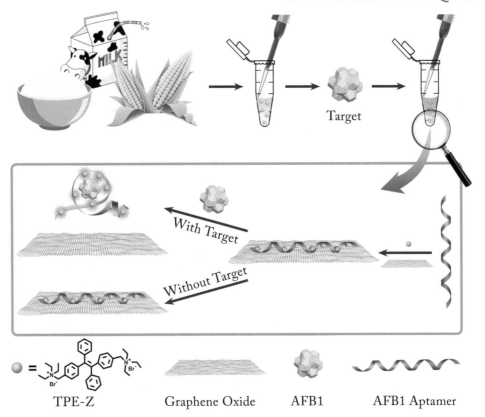

Figure 6.4: Scheme of label-free aptasensor for AFB1 dectection in food samples. Reprinted from *Talanta*, vol. 198, Jia, Y., Wu, F., Liu, P., Zhou, G., Yu, B., Lou, X., Xia, F., A label-free fluorescent aptasensor for the detection of Aflatoxin B1 in food samples using AIEgens and graphene oxide, pages 71–77. Copyright (2019), with permission from Elsevier.

graphene oxide (GO) with a detection limit of 0.25 ng/mL. TPE-Z (label-free fluorophore) and GO were introduced to create a low background fluorescence signal and increase photo-stability. There was almost no fluorescent signal without AFB1 due to the electron-transfer between TPE-Z and GO. Upon the addition of AFB1, AFB1 aptamer underwent a conformational change, forming AFB1/aptamer complex, and released from GO surface. Thus, the fluorescence of TPE-Z was recovered for detection [158].

CHAPTER 7

Nanotechnology and the Environment

7.1 ENVIRONMENTAL APPLICATIONS OF NANOTECHNOLOGY

Nanotechnology has great potential for developing innovative solutions to different environmental issues, such as water treatment, environmental sensing, remediation, and alternative energy sources [159]. The concept of environment includes not only air, water, and soil, but also organisms and wildlife in both natural and disturbed ecosystems [160].

When it comes to water and wastewater treatment, nanotechnology can be used to remove organic and inorganic contaminants and microbial pathogens through adsorption and/or photocatalysis, disinfection and microbial control, sensing and monitoring [161]. TiO_2, a renowned photocatalyst material, has been widely used for water purification. The mechanism of TiO_2 and other semiconductor photocatalysts is shown in Figure 7.1. Recently, nano TiO_2-loaded polysulfone membrane was studied for natural-rubber wastewater treatment [162]. TiO_2 addition enhanced the membrane's mechanical and separation properties and hydrophilicity—resulting in improved pollutant removal [162]. Nano TiO_2 is also used for air purification. One commercial product example is H&C TILES®. TiO_2's photocatalytic properties provide the tiles' self-cleaning and pollution-mitigation features when exposed to light (sunlight) [163].

According to the United States Environmental Protection Agency (EPA), remediation is defined as the cleanup or other methods used to remove or contain a toxic spill or hazardous materials from a Superfund site [165]. Take remediation of petroleum impurities from water for example, nanotechnologies involved include nano zero-valent iron (FeO), sponges, carbon nanostructures, and aerogels [166]. Aerogels are highly porous, lightweight materials with high surface areas and nanometer-scale pore sizes [167]. Interestingly, sugarcane biogases aerogel has been developed recently showing strong superhydrophobicity (based on the contact angles) and great potential for oil spill-cleaning (Figure 7.2) [168].

To provide an example relating to the remediation of contaminated soils, nanoscale zero valent iron (nZVI) can be discussed. nZVI has been used for heavy metal contaminants (such as As, Pb, Cr, and Sb) and persistent organic pollutants (such as trinitrotoluene and dichlorodiphenyltrichloroethane) removal [169]. The basis for the reaction is the corrosion of

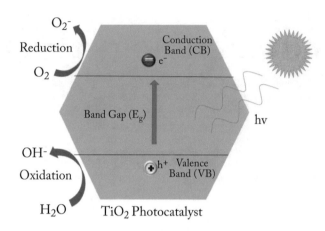

Figure 7.1: Mechanism of semiconductor photocatalyst – TiO_2. Adapted from reference [164].

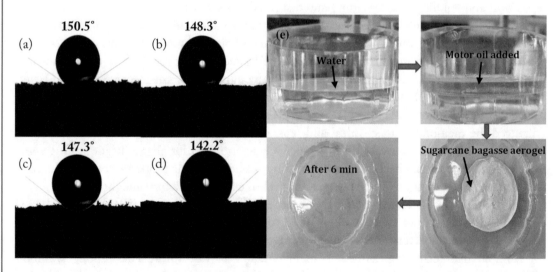

Figure 7.2: The contact angles on the external surface (a), cross-section (b), after 30 (c) and 60 (d) days exposure. Oil absorption test (e) [168]. Reprinted from *Carbohydrate Polymers*, vol. 228, Thai, Q. B., Nguyen, S. T., Ho, D. K., Tran, T. D., Huynh, D. M., Do, N. H. N., Luu, T. P., Le, P. K., Le, D. K., Phan-Thien, N. et al., Cellulose-based aerogels from sugarcane bagasse for oil spill-cleaning and heat insulation applications, 115365. Copyright (2020), with permission from Elsevier.

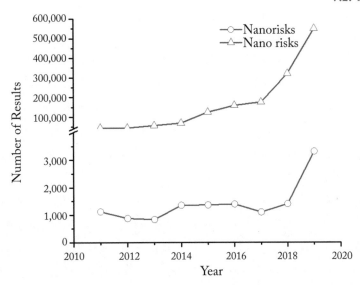

Figure 7.3: Google search results of the keywords "Nanorisks" and "Nano risks."

nZVI in the environment. Contaminants tetrachloroethane can readily accept the electrons from iron oxidation and be reduced to ethene as shown below [170]:

$$C_2 Cl_4 + 4 Fe_0 + 4 H^+ \longrightarrow C_2 H_4 + 4 Fe^{2+} + 4 Cl^-$$

7.2   NANORISKS

Google search results (2011–2019) of the keywords, "Nanorisks" and "Nano risks," have shown sharp increases in the past few years—which indicates the increasing attention and interest of investigating risks of nanotechnology (including nanomaterials) (Figure 7.3). Risk is usually defined as the probability that a hazard will occur in a given time and space and is a function of toxicity and exposure [171]. The environmental health and hazard risks associated with both nanomaterials and the applications for commercial and industrial uses are still not fully known [172].

Silver nanoparticles (Ag NPs) are considered to be one of the most commercialized nanomaterials and are used in consumer products by the companies in electronics, healthcare, cosmetics, and textiles industries [173]. Once released, Ag NPs, like other engineered nanoparticles (ENPs), can undergo multiple environmental transformations simultaneously in their natural environment [174]. Pathways and transformations of ENPs in the natural environment are summarized in Figure 7.4. It is well accepted that Ag NPs exhibit harmful effects on most organisms and even important ecosystem processes. Many studies have indicated that the toxicity of Ag NPs is not only caused by the nanoparticles themselves but also the Ag$^+$ released [175].

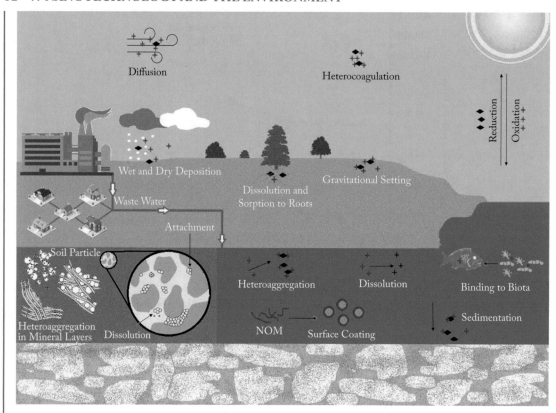

Figure 7.4: Pathways and transformations of ENPs in the natural environment. Reprinted from *Environment International*, vol. 138, Abbas, Q., Yousaf, B., Amina, Ali, M. U., Munir, M. A. M., El-Naggar, A., Rinklebe, J., Naushad, M., Transformation pathways and fate of engineered nanoparticles (ENPs) in distinct interactive environmental compartments: A review, 105646. Copyright (2020), with permission from Elsevier.

Interestingly, phytostimulation of plants exposed to Ag NPs and Ag^+ at sublethal concentrations was observed [176]. When it comes to the health impact of Ag NPs, upon exposure, *in vivo* biodistribution studies have reported Ag accumulation, local toxicity, and toxicity to distant organs [177].

7.3 RISK ASSESSMENT AND REGULATIONS

It is crucial to balance the rewards and risks of nanotechnology. Among different frameworks for characterization of environmental risk, Life Cycle Analysis (LCA) is a class of approaches that follow a product over its life stages, including (a) material acquisition and purification, (b) manufacturing and fabrication, (c) commercial uses, and (d) end-of-life product management [178].

It is an integral part of the ISO 14040:2006 [179]. However, LCA is not specific for nanomaterials/nanotechnology. There are different nano-specific frameworks, such as NanoRisk and Nano LCRA. NanoRisk framework was developed by the Environmental Defense Fund and DuPont. It is qualitative which does not generate specific guidance regarding quantitative estimation of risk associated with nanomaterials/nanotechnology [178].

Nano LCRA is an adaptive screening-level life cycle risk-assessment framework for nanotechnology. This framework contains three unique components: adaptive management for nanomaterials, life cycle thinking, and screening-level risk assessment [180]. Nano LCRA framework has been applied to assess cellulose nanomaterials [181], and it involves ten steps.

1. Describe the life cycle of the product.

2. Identify the materials and assess potential hazards in each life cycle stage.

3. Conduct an exposure assessment for each life cycle stage.

4. Identify stages of the life cycle where exposure may occur.

5. Evaluate potential human and nonhuman toxicity at key life cycle stages.

6. Analyze risk potential for the selected life cycle stages.

7. Identify key uncertainties and data gaps and communicate findings.

8. Develop mitigation/risk-management strategies.

9. Gather additional information (e.g., missing data identified from the assessment).

10. Iterate process, revisit assumptions, and adjust evaluation and management steps [178, 180].

Many government agencies are committed to promoting the responsible development of nanotechnology. Some examples (regulations and guidance) are shown in Table 7.1.

Table 7.1: Regulations and guidance pertinent to nanotechnology

Agency	Resource(s)
EU	Regulatory Aspects of Nanomaterials in the EU [182]
U.S. CDC NIOSH	Nanotechnology – Guidance & Publications [183]
USDA	Policy memorandum clarifies the status of nanotechnology in organic production and handling [184]
U.S. EPA	Control of Nanoscale Materials under the Toxic Substances Control Act [185]; Technical Fact Sheet—Nanomaterials [186]; U.S. Environmental Protection Agency Nanotechnology White Paper [187]
U.S. FDA	Nanotechnology Guidance Documents [188]
U.S. NNI	NNI 2011 Environmental, Health, and Safety (EHS) Research Strategy [189]

APPENDIX A

List of Nanobiotechnology and Biotechnology Companies

Table A.1: Nanobiotechnology and biotechnology companies in the U.S. (*Continues.*)

Name	Nanotechnology(ies)	Website
Advion BioSciences	Triversa NanoMate®: nanoelectrospray ionization technology	https://www.advion.com/
Akonni Biosystems	3D gel drop microarray technology	https://akonni.com/
Altogen Biosystems	Nanoparticle, PEG-liposome, and lipid-based transfection agents	https://altogen.com/
ANP Technologies	Nanoencapsulation using patented polymers	https://www.anptinc.com/
Aphios	Enhanced drug delivery using phospholipid nanosomes, polymer nanospheres, and protein nanoparticles	https://aphios.com/
Arch Therapeutics	Self-assembling peptide (and peptidomimetic) barriers (SAPBs)	https://www.archtherapeutics.com/
Arrayit	Microarray platform	http://www.arrayit.com/
Arrogene	Nanopolymer-based agents	https://arrogene.com/
Arrowhead Pharmaceuticals	TRiM™ platform: Targeted RNAi Molecule	https://arrowheadpharma.com/
Artificial Cell Technologies	Multilayer polypeptide nanofilm technology	https://artificialcelltech.com/
Asklepios BioPharmaceuticals (AskBio)	Recombinant adeno-associated virus (AAV)-based gene therapies	https://www.askbio.com/

Table A.1: (*Continued.*) Nanobiotechnology and biotechnology companies in the U.S. (*Continues.*)

Biodot	Non-contact and quantitative fluid dispensing systems	https://www.biodot.com/
Bioforce Nanosciences	Nano eNabler™ system: a molecular printer and ViriChip™ system: a label-free virus detection platform	http://bioforcenano.com/
Bionano Genomics	Various nanotechnologies, such as nanochannels allows only a single linearized DNA molecule to travel through	https://bionanogenomics.com/
BioPhysics Assay Laboratory (BioPAL, Inc.)	Nanoparticles for applications in magnetic resonance imaging (MRI), magnetic separation applications, and tools for time resolved fluorescent technology	http://www.biopal.com/index.html
Celsense	Novel MRI agents (perfluorocarbon emulsion)	https://celsense.com/
Celsion	LTSL (lysolipid thermally sensitive liposome) technology	https://celsion.com/
Creative Diagnostics	Nanoparticles, microparticles, and their coatings for various applications such as in vitro diagnostics and cell separation	https://www.cd-bioparticles.com/
CytImmune	Colloidal gold nanotechnology-based cancer nanomedicine	https://cytimmune.blog/
dermaCM	Nano-lipidic particles (NLPs)	http://www.dermacm.com/index.html
Dexter Magnetic Technologies	Biomagnetic separation	https://www.dextermag.com/
Dune Sciences	Nano-engineered coatings for medical textiles and smart TEM grids	http://www.dunesciences.com/home.php
Endomimetics	Natural bionanomatrix coating technologies	http://endomimetics.com/

Table A.1: (*Continued.*) Nanobiotechnology and biotechnology companies in the U.S. (*Continues.*)

Ensemble Therapeutics	Drug discovery platform -- DNA-Programmed Chemistry™ (DPC™)	https://www.ensembletx.com/
ExonanoRNA	RNA nanoparticles	https://www.exonanorna.com/
GeneFludics	Integration of novel bionano and microfluidic technologies	http://www.genefluidics.com/
Greiner Bio-One	Magnetic 3D Cell Culture: magnetization of cells with biocompatible NanoShuttle™-PL (nanoparticles).	https://www.gbo.com/en_US.html
iCeutica	SoluMatrix™ Fine Particle Technology: submicron-sized drug particles	http://www.iceutica.com/
Imagion Biosystems	Nanobiotechnology (employ magnetic nanoparticles) with an emphasis on the early detection and localization of cancer and other human diseases	https://imagionbiosystems.com/
INanoBio	Nanobiotechnology platforms for genome sequencing and diagnostics	https://inanobio.com/
Intezyn Technologies	Polymer manufacturing and formulation methods, such as polymer micelle nanoparticles	http://intezyne.com/
Kereos	Nanoscale drug delivery system for targeted delivery of therapeutic drug compounds	http://www.kereos.com/
Keystone Nano	NanoJacket and NanoLiposome Technologies	https://www.keystonenano.com/
Labcyte	Echo acoustic droplet ejection (ADE) technology: dispense in 2.5 or 25 nL increments	https://www.labcyte.com/
LLS Health	Micro and nanotechnology for pharamaceuticals	https://lubrizolcdmo.com/
Mersana Therapeutics	Antibody drug conjugate	http://www.mersana.com/index.php

Table A.1: (*Continued.*) Nanobiotechnology and biotechnology companies in the U.S. (*Continues.*)

Mirus Bio	Non-viral-based transfection agents	https://www.mirusbio.com/
NABsys	HD-Mapping™ with solid-state nanodetectors to analyze long DNA molecules	https://nabsys.com/
Nanoaffix	Graphene-based field-effect transistor platform	http://www.nanoaffix.com/
Nanobiosym	Gene-RADAR®: a portable nanotechnology platform that can rapidly and accurately detect genetic fingerprints	http://www.nanobiosym.com/
nanoComposix	Engineered nanoparticles	https://nanocomposix.com/
Nanofiber Solutions	3-D nanofiber scaffolds mimicking extracellular matrix	https://nanofibersolutions.com/
NanoHybrids	Silica-coated gold nanoparticles	https://nanohybrids.net/
NanoLight Technology	Bioluminescence reagents: luciferins (NanoFuel®s), recombinant bioluminescent proteins luciferases (NanoLight®s), photoproteins (NanoFlash™), and fluorescent proteins (NanoFluors™)	https://nanolight.com/index.php?Sid=1636255&SidC=37698598d65ee57b67600e7212a294c1
NanoMedical Systems	Silicon-based medical nanotechnology products	http://nanomedsys.com/
Nanopoint	Micro-capillary flow technology and optical cell containment device	http://www.nanopointimaging.com/
Nanoprobes	Labelling technology using metal clusters and nanoparticles as labels	http://www.nanoprobes.com/
Nanospectra Biosciences	AuroShell® particles consist of a gold metal shell and a non-conducting silica core serving as the exogenous absorber of the near-infrared laser energy delivered by the probe	https://nanospectra.com/
NanoString Technologies	A digital molecular barcoding technology	https://www.nanostring.com/

Table A.1: (*Continued.*) Nanobiotechnology and biotechnology companies in the U.S. (*Continues.*)

NanoViricides	Nanotechnology-based biomimetic anti-viral medicines	http://www.nanoviricides.com/
Novavax	Recombinant nanoparticle vaccine technology	https://www.novavax.com/
Parabon NanoaLabs	Engineering DNA for next-generation therapeutics and forensics	https://parabon-nanolabs.com/
PDS Biotechnology	Versamune® nanotechnology platform: based on novel and structurally specific synthetic and positively charged lipids	http://www.pdsbiotech.com/
PharmaIN	PGC™: a novel drug carrier	https://pharmain.com/
Pharma Seq	p-Chip®, the world's first and only ultra-small (500 microns on a side) microtransponder	http://www.pharmaseq.com/
Phoenix S&T	Nanospray for proteomics and biomarker discovery applications.	https://phoenix-st.com/
Platypus Technologies	Nano-structured surfaces	https://www.platypustech.com/
PolyMicrospheres	Uniform polymer microspheres for clinical diagnostics, and microsphere- and nanoparticle-based drug delivery systems	https://www.polymicrospheres.com/
Quanterix	Simoa™ (Single Molecule Array) Technology	https://www.quanterix.com/
Selecta Biosciences	ImmTOR selective immune technology (nanoparticle-based system)	https://www.selectabio.com/
Sirnaomics	Polypeptide Nano-Particle (PNP) technology for small interfering RNA (siRNA) drug delivery	https://sirnaomics.com/
Uluru	Altrazeal® platform: Flexible HydrogelNanoparticle Wound Dressing (flakes of freeze-dried HEMA and HPMA polymers)	https://www.uluruinc.info/

Table A.1: (*Continued.*) Nanobiotechnology and biotechnology companies in the U.S.

Vista Therapeutics	Silicon nanowire FETs for biomarkers detection	http://www.vistananobiosciences.com/
Zylö Therapeutics	Z-pod™ delivery technology (xerogel-derived particles)	https://www.zylotherapeutics.com/

APPENDIX B

List of Nanotechnology Research Centers in the U.S.

Table B.1: Nanotechnology Research Centers in the U.S. (*Continues.*)

Name	Website
Bio- and Nano-Technology Center	https://www.pharmacy.umaryland.edu/centers/bio-and-nano-technology-center/
Birck Nanotechnology Center	https://www.purdue.edu/discoverypark/birck/
California NanoSystems Institute	https://cnsi.ucla.edu/
Center for Advanced Materials and Nanotechnology	https://www.lehigh.edu/nano/
Center for Affordable Nanoengineering of Polymeric Biomedical Devices	https://nsec.osu.edu/
Center for Cancer Nanotechnology Excellence for Translational Diagnostics	https://med.stanford.edu/ccne.html
Center for Drug Delivery and Nanomedicine	http://cddn.unmc.edu/template_view.cfm?PageID=29
Center for Enhanced Nanofluidic Transport	https://cent.mit.edu/
Center for the Environmental Implications of NanoTechnology	http://ceint.duke.edu/
Center for Environmental Nanoscience and Risk	https://www.sc.edu/study/colleges_schools/public_health/research/research_centers/center_for_environmental_nanoscience_and_risk/index.php
Center for Nanomedicine and Engineering,	http://nanomedicine.ucsd.edu/
Center for High-rate Nanomanufacturing	https://www.nanomanufacturing.us/
Center for Integrative Nanotechnology Sciences	https://ualr.edu/nanotechnology/
Center for Magnetism and Magnetic Nanostructures	https://www.uccs.edu/physics/magnetism_and_magnetic

Table B.1: (*Continued.*) Nanotechnology Research Centers in the U.S. (*Continues.*)

Center for Nanomedicine, Johns Hopkins School of Medicine	https://cnm-hopkins.org/
Center for Nanomedicine, University of California Santa Barbara	https://www.cnm.ucsb.edu/
Center for Nanoscale Science and Technology, National Institute of Standards and Technology (NTSI)	https://www.nist.gov/cnst
Center for Nanoscale Systems	https://cns1.rc.fas.harvard.edu/
Center for Nano-Science and Technology, Texas A&M University	http://cnst.tamu.edu/
Center for Nano Science and Technology, University of Notre Dame	https://nano.nd.edu/
Center for Nanoscience, University of Missouri-St. Louis	http://www.umsl.edu/services/ora/cns/
Center for Nanotechnology in Drug Delivery	https://pharmacy.unc.edu/research/centers/cndd/
Center for Nanotechnology and Molecular Materials	https://www.sun.ictas.vt.edu/
Center for Nanotechnology and Nanotoxicology	https://www.hsph.harvard.edu/nano/
Center for Nanotechnology in Society	http://cns.asu.edu/
Center for Pharmaceutical Biotechnology and Nanomedicine	https://bouve.northeastern.edu/cpbn/
Center for Probing the Nanoscale	https://web.stanford.edu/group/cpn/
Center for Sustainable Nanotechnology	https://susnano.wisc.edu/team/
Clemson Nanomaterials Institute	http://www.cniatclemson.net/
Columbia Nano Initiative	https://cni.columbia.edu/
ElectroOptics Research Institute and Nanotechnology Center	http://eri.louisville.edu/
Genetically Engineered Materials and Micro/Nano Devices	http://www.gems.gatech.edu/index.html
George J. Kostas Nanoscale Technology and Manufacturing Research Center	http://kostas.neu.edu/

Table B.1: (*Continued.*) Nanotechnology Research Centers in the U.S. (*Continues.*)

Institute for Electronics and Nanotechnology	http://ien.gatech.edu/
Institute for Molecular Manufacturing	http://www.imm.org/
Institute for Molecular and Nanoscale Innovation	https://www.brown.edu/research/institute-molecular-nanoscale-innovation/
Institute for Nano-Engineered Systems	https://www.nano.uw.edu/
Institute for Nanoscale and Quantum Scientific and Technological Advanced Research	https://nanostar.virginia.edu/
Institute for Soldier Nanotechnologies	http://isn.mit.edu/
Interdisciplinary Center for Nanotoxicity	http://icnanotox.org/research/
Integrative NanoScience Institute	https://insi.fsu.edu/
International Institute for Nanotechnology	https://www.iinano.org/
Johns Hopkins Institute for NanoBioTechnology	https://inbt.jhu.edu/
Kavli Energy NanoScience Institute	https://kavli.berkeley.edu/
Kavli Institute at Cornell for Nanoscale Science	http://www.kicnano.cornell.edu/
Kavli Institute for Bionano Science and Technology	https://kavli.seas.harvard.edu/home
Kavli Nanoscience Institute, California Institute of Technology	http://www.kni.caltech.edu/
Alan G. MacDiarmid NanoTech Institute	https://centers.utdallas.edu/nanotech/
Marble Center for Cancer Nanomedicine	http://nanomedicine.mit.edu/
Maryland Nanocenter	https://www.nanocenter.umd.edu/
Mason Nanotechnology Initiative	http://nano.gmu.edu/index.html
MassNanoTech Institute	http://www.umass.edu/massnanotech/
MSK-Cornell Center for Translation of Cancer Nanomedicine	https://www.mskcc.org/research-programs/translation-cancer-nanomedicine
Micro/Nano Technology Center	https://louisville.edu/micronano
Minnesota Nano Center	https://www.mnc.umn.edu/
Nano Institute of Utah	https://nanoinstitute.utah.edu/
Nano/Bio Interface Center	http://www.nanotech.upenn.edu/

Table B.1: (*Continued.*) Nanotechnology Research Centers in the U.S. (*Continues.*)

NanoForestry, Purdue University - US Forest Service Forest Products Laboratory (FPL)	https://engineering.purdue.edu/nanotrees/index.shtml
Nanomanufacturing Center	https://www.uml.edu/research/nano/
Nanoscale Science and Engineering Center	https://nsec.wisc.edu/
Nanoscale Science and Engineering Center for Directed Assembly of Nanostructures	http://www.rpi.edu/dept/nsec/
Nanoscience Institute for Medical and Engineering Technology	https://www.eng.ufl.edu/nimet/
Nanoscience and Nanotechnology Institute	https://nanotech.uiowa.edu/
Nanosystems Engineering Research Center for Nanotechnology-Enabled Water Treatment	http://www.newtcenter.org/
Nanotechnology Characterization Laboratory, National Cancer Institute	https://ncl.cancer.gov/
Nanotechnology Innovation Center, Boston University	https://www.bu.edu/nano-bu/
Nanotechnology Innovation Center of Kansas State	https://nicks.ksu.edu/
Nanotechnology Research Center, National Institute for Occupational Safety and Health (NIOSH)	https://www.cdc.gov/niosh/programs/nano/default.html
Nanotechnology Research and Education Center	http://nnrc.eng.usf.edu/
Nanovaccine Institute	https://www.nanovaccine.iastate.edu/
Nanoworld Laboratories	http://milkyway.mie.uc.edu/nanoworldsmart
Nebraska Center for Materials and Nanoscience	https://ncmn.unl.edu/
Nevada Nanotechnology Center	http://nanotechnology.unlv.edu/
North Carolina Center of Innovation Network	https://nccoin.org/
North Carolina Center for Nanoscale Materials	https://users.physics.unc.edu/~zhou/muri/
Gertrude E. and John M. Petersen Institute of NanoScience and Engineering	http://www.nano.pitt.edu/
Rensselaer Nanotechnology Center	http://rnc.rpi.edu/

Table B.1: (*Continued.*) Nanotechnology Research Centers in the U.S.

Research in the Nanomaterials Group	https://nano.materials.drexel.edu/
Science of Nanoscale Systems and their Device Applications	http://www.nsec.harvard.edu/
Shimadzu Institute Nano Technology Research Center	http://www.uta.edu/sirt/nano/
Singh Center for Nanotechnology	https://www.nano.upenn.edu/
Vanderbilt Institute of Nanoscale Science and Engineering	https://www.vanderbilt.edu/vinse/
Virginia Tech Center for Sustainable Nanotechnology	https://www.sun.ictas.vt.edu/
Wisconsin Center for NanoBioSystems	https://pharmacy.wisc.edu/centers/wiscnano/
Wisconsin Centers for Nanoscale Technology	https://wcnt.wisc.edu/
Zyvex Labs	https://www.zyvexlabs.com/

APPENDIX C

List of Nanotechnology Journals

Table C.1: List of nanotechnology journals (*Continues.*)

Name	Website
ACS Nano	https://pubs.acs.org/journal/ancac3
Advances in Nano Research	http://www.techno-press.org/?journal=anr&-subpage=1
Applied Nanoscience	https://link.springer.com/journal/13204
Artificial Cells, Nanomedicine, and Biotechnology: An International Journal	https://www.tandfonline.com/loi/ianb20
Beilstein Journal of Nanotechnology	https://www.beilstein-journals.org/bjnano/home
Bioinspired, Biomimetic and Nanobiomaterials	https://www.icevirtuallibrary.com/toc/jbibn/current
ChemNanoMat	https://onlinelibrary.wiley.com/journal/2199692x
Current Nanoscience	https://benthamscience.com/journal/index.php?journalID=cnano
Digest Journal of Nanomaterials and Biostructures	http://www.chalcogen.ro/index.php/journals/digest-journal-of-nanomaterials-and-biostructures
IEEE Transactions on NanoBioscience	https://ieeexplore.ieee.org/xpl/RecentIssue.jsp?punumber=7728
IEEE Transactions on Nanotechnology	https://ieeexplore.ieee.org/xpl/RecentIssue.jsp?punumber=7729
IET Nanobiotechnology	https://digital-library.theiet.org/content/journals/iet-nbt
International Journal of Nanomedicine	https://www.dovepress.com/international-journal-of-nanomedicine-journal

Table C.1: (*Continued.*) List of nanotechnology journals (*Continues.*)

International Journal of Nanotechnology	https://www.inderscience.com/jhome.php?-jcode=ijnt&csid=fevmdn4fnksa8g5ebpbta-hods0
Journal of Biomedical Nanotechnology	http://www.aspbs.com/jnn/
Journal of Micro/Nanolithography, MEMS, and MOEMS	https://www.spiedigitallibrary.org/jour-nals/journal-of-micro-nanolithogra-phy-mems-and-moems?SSO=1
Journal of Nanobiotechnology	https://jnanobiotechnology.biomedcentral.com/
Journal of Nanomaterials	https://www.hindawi.com/journals/jnm/
Journal of Nanoparticle Research	https://www.springer.com/journal/11051
Journal of Nano Research	https://www.scientific.net/JNanoR/Details
Journal of Nanoscience and Nanotechnology	http://www.aspbs.com/jnn/
Micro and Nano Letters	https://digital-library.theiet.org/content/jour-nals/mnl
Microsystems and Nanoengineering	https://www.nature.com/micronano/
Nano	https://www.worldscientific.com/worldscinet/nano
Nano Energy	https://www.journals.elsevier.com/nano-en-ergy/
Nano Letters	https://pubs.acs.org/journal/nalefd
Nanomaterials	https://www.mdpi.com/journal/nanomaterials
Nanomaterials and Nanotechnology	https://journals.sagepub.com/home/nax
Nanomedicine	https://www.futuremedicine.com/loi/nnm
Nanomedicine: Nanotechnology Biology and Medicine	https://www.sciencedirect.com/journal/nano-medicine-nanotechnology-biology-and-med-icine
Nano-Micro Letters	http://www.nmletters.org/
Nano Research	https://www.springer.com/journal/12274
Nanoscale	https://pubs.rsc.org/en/journals/journalissues/nr#!recentarticles&adv
Nanoscale Horizons	https://www.rsc.org/journals-books-databases/about-journals/nanoscale-horizons/

Table C.1: (*Continued.*) List of nanotechnology journals

Nanoscale Research Letters	https://nanoscalereslett.springeropen.com/
Nanoscience and Nanotechnology Letters	http://www.aspbs.com/nnl.htm
Nanotechnology	https://iopscience.iop.org/journal/0957-4484
Nanotechnology Reviews	https://www.degruyter.com/view/j/ntrev
Nano Today	https://www.journals.elsevier.com/nano-today/
Nature Nanotechnology	https://www.nature.com/nnano/
Precision Engineering—Journal of the International Societies for Precision Engineering and Nanotechnology	https://www.journals.elsevier.com/precision-engineering/
Recent Patents on Nanotechnology	https://benthamscience.com/journal/index.php?journalID=rpnanotec
Wiley Interdisciplinary Reviews-Nanomedicine and Nanobiotechnology	https://onlinelibrary.wiley.com/journal/19390041
*** This is by no means an exhaustive list and is subject to updates and additions.	

APPENDIX D

Determining Nanoparticle Size Using XRD Data

The Debye-Scherrer equation (shown below) is the key to determining the nanoparticle size. FWHM is shown in Figure D.1, which can be determined manually by software like NIH Image J.

Debye-Scherrer equation: $\quad D = K\lambda/\beta \cos\theta,$

D: crystal size,

K: Scherrer constant ($\sim$ 1 for spherical particles),

λ: wavelength of X-ray (0.15418 nm; mostly used),

θ: Braggs angle,

β: full width at half maximum (FWHM) of the peak.

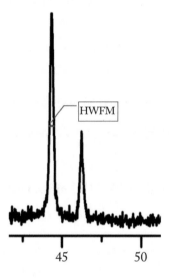

Figure D.1: A screenshot showing FWHM of a peak of an XRD graph (x-axis: 2θ (degree) and y-axis: intensity). The length is determined to be 0.331 (in response to axis-scale).

APPENDIX E

Different Nanomaterials/Systems/Structures

Table E.1: **Different nanomaterials/systems/structures**

0 D	Nanoparticles, nanocapsules, nanocrystals including quantum dots, nanoclusters, nanospheres, etc.
1 D	Nanorods, nanofibers, nanobelts, nanotubes, nanowires, etc.
2 D	Nanofilm, nanosheets, nanoflakes, nanoplates, etc.
****Based on how many dimensions are not at the nano-scale.	

APPENDIX F

Three-Dimensional Structures of Virus-Like Particles

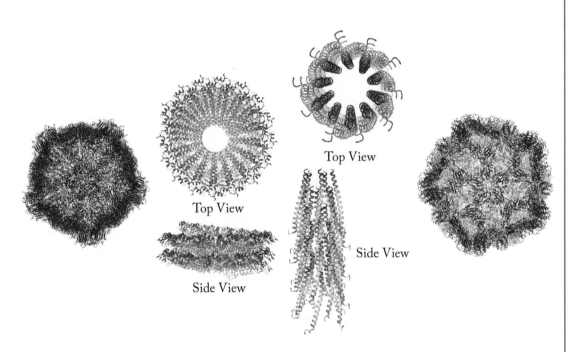

Figure F.1: Three-dimensional structures of, from left to right, empty Cowpea Mosaic Virus (eCPMV; PDB ID: 5FMO), Tobacco Mosaic Virus capsid (TMV; PDB ID: 6R7M), M13 bacteriophage capsid (PDB ID: 2MJZ), and Macrobrachium rosenbergii nodavirus VLP (MrNV; PDB ID: 6JJD). Images of 5FMO (N. T. Huynh, E.L. Hesketh, P. Saxena, Y. Meshcheriakova, Y. Ku, L.T. Hoang, J. E. Johnson, N.A. Ranson, G. P. Lomonossoff, V. S. Reddy (2016) Crystal Structure and Proteomics Analysis of Empty Virus-like Particles of Cowpea Mosaic Virus. *Structure* 24(4): 567-575), 6R7M (C. Schmidli, S. Albiez, L. Rima, R. Righetto, I. Mohammed, P. Oliva, L. Kovacik, H. Stahlberg, T. Braun (2019) *Proceedings of the National Academy of Sciences* 116 (30): 15007-15012), and 2MJZ (O. Morag, N.G. Sgourakis, D. Baker, A. Goldbourt

(2015) The NMR-Rosetta capsid model of M13 bacteriophage reveals a quadrupled hydrophobic packing epitope. *Proceedings of the National Academy of Sciences of the United States of America* 112(4), 971–976) created with NGL (A.S. Rose, A.R. Bradley, Y. Valasatava, J.D. Duarte, A. Prlić, P.W. Rose (2018) NGL viewer: web-based molecular graphics for large complexes. *Bioinformatics* 34: 3755–3758) and RCSB Protein Data Bank (www.pdb.org).

Bibliography

[1] What is Nanotechnology? https://www.nano.gov/nanotech-101/what/definition 1

[2] Nanoscience and nanotechnologies: Opportunities and uncertainties, *The Royal Society and the Royal Academy of Engineering*, 2004. https://royalsociety.org/-/media/Royal_Society_Content/policy/publications/2004/9693.pdf 1

[3] The appropriateness of existing methodologies to assess the potential risks associated with engineered and adventitious products of nanotechnologies, *SCENIHR, European Commission*, 2006. https://ec.europa.eu/health/ph_risk/committees/04_scenihr/docs/scenihr_o_003b.pdf 1

[4] Feynman, R. There's plenty of room at the bottom. *Engineering and Science*, pages 22–36, 1960. 1

[5] Ashby, M. F., Ferreira, P. J., and Schodek, D. L. Chapter 2—An evolutionary perspective. *Nanomaterials, Nanotechnologies and Design*, Ashby, M. F., Ferreira, P. J., and Schodek, D. L., Eds., Butterworth–Heinemann, Boston, 2009. DOI: 10.1016/b978-0-7506-8149-0.x0001-3. 3

[6] Nanotechnology Timeline. https://www.nano.gov/timeline 2, 5

[7] Shirai, Y., Osgood, A. J., Zhao, Y., Kelly, K. F., and Tour, J. M. Directional control in thermally driven single-molecule nanocars. *Nano Letters*, 5:2330–2334, 2005. DOI: 10.1021/nl051915k. 4

[8] College of Nanoscience and Engineering, SUNY Polytechnic Institute. https://sunypoly.edu/academics/colleges/college-nanoscale-science-engineering.html 5

[9] Nasrollahzadeh, M., Issaabadi, Z., Sajjadi, M., Sajadi, S. M., and Atarod, M. Chapter 2—Types of nanostructures. *Interface Science and Technology*, Nasrollahzadeh, M., Sajadi, S. M., Sajjadi, M., Issaabadi, Z., and Atarod, M., Eds., Elsevier, 2019. DOI: 10.1016/B978-0-12-813586-0.00002-X. 7

[10] Nicholson, E. D. The ancient craft of gold beating. *Gold Bulletin*, 12:161–166, 1979. DOI: 10.1007/bf03215119. 7

[11] Raval, J. P., Joshi, P., and Chejara, D. R. Chapter 9—Carbon nanotube for targeted drug delivery. *Applications of Nanocomposite Materials in Drug Delivery*, Inamuddin, Asiri, A.

M., and Mohammad, A., Eds., Woodhead Publishing, 2018. DOI: 10.1016/B978-0-12-813741-3.00009-1. 7

[12] United States Patent: Process for producing an anodic aluminum oxide membrane (Inventor: Smith, A.), 1974. 7

[13] Gosai, A., Hau Yeah, B. S., Nilsen-Hamilton, M., and Shrotriya, P. Label free thrombin detection in presence of high concentration of albumin using an aptamer-functionalized nanoporous membrane. *Biosensors and Bioelectronics*, 126:88–95, 2019. DOI: 10.1016/j.bios.2018.10.010. 7

[14] Shi, X., Zhou, W., Ma, D., Ma, Q., Bridges, D., Ma, Y. and Hu, A. Electrospinning of nanofibers and their applications for energy devices. *Journal of Nanomaterials*, 20, 2015. DOI: 10.1155/2015/140716. 9

[15] Xue, J., Wu, T., Dai, Y., and Xia, Y. Electrospinning and electrospun nanofibers: Methods, materials, and applications. *Chemical Reviews*, 119:5298–5415, 2019. DOI: 10.1021/acs.chemrev.8b00593. 9

[16] Xue, J., Xie, J., Liu, W., and Xia, Y. Electrospun nanofibers: New concepts, materials, and applications. *Accounts of Chemical Research*, 50:1976–1987, 2017. DOI: 10.1021/acs.accounts.7b00218. 8, 9

[17] Cherwin, A., Namen, S., Rapacz, J., Kusik, G., Anderson, A., Wang, Y., Kaltchev, M., Schroeder, R., O'Connell, K., Stephens, S., et al. Design of a novel oxygen therapeutic using polymeric hydrogel microcapsules mimicking red blood cells. *Pharmaceutics*, 11(583), 2019. DOI: 10.3390/pharmaceutics11110583. 9, 11

[18] Luo, C. J. and Edirisinghe, M. Core-liquid-induced transition from coaxial electrospray to electrospinning of low-viscosity poly(lactide-co-glycolide) sheath solution. *Macromolecules*, 47:7930–7938, 2014. DOI: 10.1021/ma5016616. 10

[19] Iqbal, S., Rashid, M. H., Arbab, A. S., and Khan, M. Encapsulation of anticancer drugs (5-fluorouracil and paclitaxel) into polycaprolactone (PCL) nanofibers and in vitro testing for sustained and targeted therapy. *Journal of Biomedical Nanotechnology*, 13:355–366, 2017. DOI: 10.1166/jbn.2017.2353. 10

[20] Jayasinghe, S. N. Biojets in regenerative biology and medicine. *Materials Today*, 14:202–211, 2011. DOI: 10.1016/s1369-7021(11)70115-8. 10

[21] Hwang, Y. J., Choi, S., and Kim, H. S. Structural deformation of PVDF nanoweb due to electrospinning behavior affected by solvent ratio. *e-Polymers*, 18(339), 2018. DOI: 10.1515/epoly-2018-0037. 10

[22] Alfaro De Prá, M. A., Ribeiro-do-Valle, R. M., Maraschin, M., and Veleirinho, B. Effect of collector design on the morphological properties of polycaprolactone electrospun fibers. *Materials Letters*, 193:154–157, 2017. DOI: 10.1016/j.matlet.2017.01.102. 11

[23] Salehi-Nik, N., Rezai Rad, M., Nazeman, P., and Khojasteh, A. Chapter 3—Polymers for oral and dental tissue engineering. *Biomaterials for Oral and Dental Tissue Engineering*, Tayebi, L. and Moharamzadeh, K., Eds., Woodhead Publishing, 2017. DOI: 10.1016/B978-0-08-100961-1.00003-7. 10

[24] Stealey, S., Guo, X., Majewski, R., Dyble, A., Lehman, K., Wedemeyer, M., Steeber, D. A., Kaltchev, M. G., Chen, J., and Zhang, W. Calcium-oligochitosan-pectin microcarrier for colonic drug delivery. *Pharmaceutical Development and Technology*, 25:260–265, 2019. DOI: 10.1080/10837450.2019.1691591. 11

[25] Stealey, S., Guo, X., Ren, L., Bryant, E., Kaltchev, M., Chen, J., Kumpaty, S., Hua, X., and Zhang, W. Stability improvement and characterization of bioprinted pectin-based scaffold. *Journal of Applied Biomaterials and Functional Materials* 17, 2019. DOI: 10.1177/2280800018807108. 11

[26] Banks, A., Guo, X., Chen, J., Kumpaty, S., and Zhang, W. Novel bioprinting method using a pectin based bioink. *Technology and Health Care* 25:651–655, 2017. DOI: 10.3233/thc-160764. 11

[27] Crouse, J. Z., Mahuta, K. M., Mikulski, B. A., Harvestine, J. N., Guo, X., Lee, J. C., Kaltchev, M. G., Midelfort, K. S., Tritt, C. S., Chen, J., et al. Development of a microscale red blood cell-shaped pectin-oligochitosan hydrogel system using an electrospray-vibration method: Preparation and characterization. *Journal of Applied Biomaterials and Functional Materials*, 13:e326–331, 2015. DOI: 10.5301/jabfm.5000250. 11

[28] Zhang, W., Mahuta, K. M., Mikulski, B. A., Harvestine, J. N., Crouse, J. Z., Lee, J. C., Kaltchev, M. G., and Tritt, C. S. Novel pectin-based carriers for colonic drug delivery. *Pharmaceutical Development and Technology*, 21:127–130, 2016. DOI: 10.3109/10837450.2014.965327. 11

[29] Johnson, D. L., Ziemba, R. M., Shebesta, J. H., Lipscomb, J. C., Wang, Y., Wu, Y., O'Connell, K. D., Kaltchev, M. G., van Groningen, A., Chen, J., et al. Design of pectin-based bioink containing bioactive agent-loaded microspheres for bioprinting. *Biomedical Physics and Engineering Express*, 5:067004, 2019. DOI: 10.1088/2057-1976/ab4dbc. 11

[30] Zhang, W., Bissen, M. J., Savela, E. S., Clausen, J. N., Fredricks, S. J., Guo, X., Paquin, Z. R., Dohn, R. P., Pavelich, I. J., Polovchak, A. L., et al. Design of artificial red blood cells using polymeric hydrogel microcapsules: Hydrogel stability improvement and

polymer selection. *The International Journal of Artificial Organs*, 39:518–523, 2016. DOI: 10.5301/ijao.5000532. 11

[31] McCune, D., Guo, X., Shi, T., Stealey, S., Antrobus, R., Kaltchev, M., Chen, J., Kumpaty, S., Hua, X., Ren, W., et al. Electrospinning pectin-based nanofibers: A parametric and cross-linker study. *Applied Nanoscience*, 8:33–40, 2018. DOI: 10.1007/s13204-018-0649-4. 11, 12

[32] Cui, S., Yao, B., Sun, X., Hu, J., Zhou, Y., and Liu, Y. Reducing the content of carrier polymer in pectin nanofibers by electrospinning at low loading followed with selective washing. *Materials Science and Engineering C*, 59:885–893, 2016. DOI: 10.1016/j.msec.2015.10.086. 11, 12

[33] Cui, S., Yao, B., Gao, M., Sun, X., Gou, D., Hu, J., Zhou, Y., and Liu, Y. Effects of pectin structure and crosslinking method on the properties of crosslinked pectin nanofibers. *Carbohydrate Polymers*, 157:766–774, 2017. DOI: 10.1016/j.carbpol.2016.10.052. 13

[34] Hamouda, R. A., Hussein, M. H., Abo-elmagd, R. A., and Bawazir, S. S. Synthesis and biological characterization of silver nanoparticles derived from the cyanobacterium Oscillatoria limnetica. *Scientific Reports*, 9:13071, 2019. DOI: 10.1038/s41598-019-49444-y. 15

[35] Firdhouse, M. J. and Lalitha, P. Biosynthesis of silver nanoparticles and its applications. *Journal of Nanotechnology*, 18, 2015. DOI: 10.1155/2015/829526. 15

[36] Bulavinets, T., Varyshchuk, V., Yaremchuk, I., and Bobitski, Y. Design and synthesis of silver nanoparticles with different shapes under the influence of photon flows. *Nanooptics, Nanophotonics, Nanostructures, and their Applications*, Fesenko, O. and Yatsenko, L., Eds., Springer, 2018. DOI: 10.1007/978-3-319-91083-3_16. 15, 16

[37] Yeh, Y.-C., Creran, B., and Rotello, V. M. Gold nanoparticles: Preparation, properties, and applications in bionanotechnology. *Nanoscale*, 4:1871–1880, 2012. DOI: 10.1039/c1nr11188d. 15

[38] Lee, J.-H., Cho, H.-Y., Choi, H. K., Lee, J.-Y., and Choi, J.-W. Application of gold nanoparticle to plasmonic biosensors. *International Journal of Molecular Sciences*, 19:2021, 2018. DOI: 10.3390/ijms19072021. 15

[39] Ma, N., Wu, F.-G., Zhang, X., Jiang, Y.-W., Jia, H.-R., Wang, H.-Y., Li, Y.-H., Liu, P., Gu, N., and Chen, Z. Shape-dependent radiosensitization effect of gold nanostructures in cancer radiotherapy: Comparison of gold nanoparticles, nanospikes, and nanorods. *ACS Applied Materials and Interfaces*, 9:13037–13048, 2017. DOI: 10.1021/acsami.7b01112. 15

[40] Dreaden, E. C., Alkilany, A. M., Huang, X., Murphy, C. J., and El-Sayed, M. A. The golden age: Gold nanoparticles for biomedicine. *Chemical Society Reviews*, 41:2740–2779, 2012. DOI: 10.1039/C1CS15237H. 15

[41] Iravani, S., Korbekandi, H., Mirmohammadi, S. V., and Zolfaghari, B. Synthesis of silver nanoparticles: Chemical, physical and biological methods. *Research in Pharmaceutical Sciences*, 9:385–406, 2014. 15

[42] Sun, Y. and Xia, Y. Shape-controlled synthesis of gold and silver nanoparticles. *Science*, 298:2176–2179, 2002. DOI: 10.1126/science.1077229. 16

[43] Suchomel, P., Kvitek, L., Prucek, R., Panacek, A., Halder, A., Vajda, S., and Zboril, R. Simple size-controlled synthesis of Au nanoparticles and their size-dependent catalytic activity. *Scientific Reports*, 8:4589, 2018. DOI: 10.1038/s41598-018-22976-5. 16

[44] Raza, M. A., Kanwal, Z., Rauf, A., Sabri, A. N., Riaz, S., and Naseem, S. Size- and shape-dependent antibacterial studies of silver nanoparticles synthesized by wet chemical routes. *Nanomaterials*, 6:74, 2016. DOI: 10.3390/nano6040074. 16

[45] Piñero, S., Camero, S., and Blanco, S. Silver nanoparticles: Influence of the temperature synthesis on the particles' morphology. *Journal of Physics: Conference Series*, 786:012020, 2017. DOI: 10.1088/1742-6596/786/1/012020. 16

[46] Samanta, S., Agarwal, S., Nair, K. K., Harris, R. A., and Swart, H. Biomolecular assisted synthesis and mechanism of silver and gold nanoparticles. *Materials Research Express*, 6:082009, 2019. DOI: 10.1088/2053-1591/ab296b. 17

[47] Johnson, D. L., Wang, Y., Stealey, S. T., Alexander, A. K., Kaltchev, M. G., Chen, J., and Zhang, W. Biosynthesis of silver nanoparticles using upland cress: Purification, characterisation, and antimicrobial activity. *Micro and Nano Letters*, 15:110–113, 2020. DOI: 10.1049/mnl.2019.0528. 17, 18, 20, 21

[48] Ovais, M., Khalil, A. T., Ayaz, M., Ahmad, I., Nethi, S. K., and Mukherjee, S. Biosynthesis of metal nanoparticles via microbial enzymes: A mechanistic approach. *International Journal of Molecular Sciences*, 19:4100, 2018. DOI: 10.3390/ijms19124100. 17

[49] Lin, W., Zhang, W., Zhao, X., Roberts, A. P., Paterson, G. A., Bazylinski, D. A., and Pan, Y. Genomic expansion of magnetotactic bacteria reveals an early common origin of magnetotaxis with lineage-specific evolution. *The ISME Journal*, 12:1508–1519, 2018. DOI: 10.1038/s41396-018-0098-9. 18

[50] Stürzenbaum, S. R., Höckner, M., Panneerselvam, A., Levitt, J., Bouillard, J. S., Taniguchi, S., Dailey, L. A., Khanbeigi, R. A., Rosca, E. V., Thanou, M., et al. Biosynthesis of luminescent quantum dots in an earthworm. *Nature Nanotechnology*, 8:57–60, 2013. DOI: 10.1038/nnano.2012.232. 18

[51] Ginet, N., Pardoux, R., Adryanczyk, G., Garcia, D., Brutesco, C., and Pignol, D. Single-step production of a recyclable nanobiocatalyst for organophosphate pesticides biodegradation using functionalized bacterial magnetosomes. *PLOS ONE*, 6:e21442, 2011. DOI: 10.1371/journal.pone.0021442. 18

[52] Vilchis-Nestor, A. R., Sánchez-Mendieta, V., Camacho-López, M. A., Gómez-Espinosa, R. M., Camacho-López, M. A., and Arenas-Alatorre, J. A. Solventless synthesis and optical properties of Au and Ag nanoparticles using Camellia sinensis extract. *Materials Letters*, 62:3103–3105, 2008. DOI: 10.1016/j.matlet.2008.01.138. 18

[53] Song, J. Y. and Kim, B. S. Rapid biological synthesis of silver nanoparticles using plant leaf extracts. *Bioprocess and Biosystems Engineering*, 32:79–84, 2009. DOI: 10.1007/s00449-008-0224-6. 18

[54] Agerbirk, N. and Olsen, C. E. Isoferuloyl derivatives of five seed glucosinolates in the crucifer genus Barbarea. *Phytochemistry*, 72:610–623, 2011. DOI: 10.1016/j.phytochem.2011.01.034. 18

[55] Pedras, M. S. C., Alavi, M., and To, Q. H. Expanding the nasturlexin family: Nasturlexins C and D and their sulfoxides are phytoalexins of the crucifers Barbarea vulgaris and B. verna. *Phytochemistry*, 118:131–138, 2015. DOI: 10.1016/j.phytochem.2015.08.009. 18

[56] Xiao, Z., Rausch, S. R., Luo, Y., Sun, J., Yu, L., Wang, Q., Chen, P., Yu, L., and Stommel, J. R. Microgreens of brassicaceae: Genetic diversity of phytochemical concentrations and antioxidant capacity. *LWT—Food Science and Technology*, 101:731–737, 2019. DOI: 10.1016/j.lwt.2018.10.076. 18

[57] Qing, Y., Cheng, L., Li, R., Liu, G., Zhang, Y., Tang, X., Wang, J., Liu, H., and Qin, Y. Potential antibacterial mechanism of silver nanoparticles and the optimization of orthopedic implants by advanced modification technologies. *International Journal of Nanomedicine*, 13:3311–3327, 2018. DOI: 10.2147/ijn.s165125. 20, 22

[58] The principles of dynamic light scattering. https://wiki.anton-paar.com/us-en/the-principles-of-dynamic-light-scattering/ 20

[59] Dahman, Y., Caruso, G., Eleosida, A., and Hasnain, S. Chapter 10—Self-assembling nanostructures. *Nanotechnology and Functional Materials for Engineers*, Dahman, Y., Ed., Elsevier, 2017. DOI: 10.1016/B978-0-323-51256-5.00010-1. 23

[60] Ozin, G. A., Hou, K., Lotsch, B. V., Cademartiri, L., Puzzo, D. P., Scotognella, F., Ghadimi, A., and Thomson, J. Nanofabrication by self-assembly. *Materials Today*, 12:12–23, 2009. DOI: 10.1016/s1369-7021(09)70156-7. 23

[61] Subramani, K. and Ahmed, W. Chapter 12—Self-assembly of proteins and peptides and their applications in bionanotechnology and dentistry. *Emerging Nanotechnologies in Dentistry*, 2nd ed., Subramani, K. and Ahmed, W., Eds., William Andrew Publishing, 2018. DOI: 10.1016/B978-0-12-812291-4.00012-1. 23

[62] Wang, L., Gong, C., Yuan, X., and Wei, G. Controlling the self-assembly of biomolecules into functional nanomaterials through internal interactions and external stimulations: A review. *Nanomaterials*, 9:285, 2019. DOI: 10.3390/nano9020285. 23

[63] Arora, A. A. and de Silva, C. Beyond the smiley face: Applications of structural DNA nanotechnology. *Nano Reviews and Experiments*, 9:1430976, 2018. DOI: 10.1080/20022727.2018.1430976. 23

[64] Jiang, S. Thermodynamics and kinetics of DNA tile-based self-assembly. Arizona State University, 2016. 23

[65] Praetorius, F., Kick, B., Behler, K. L., Honemann, M. N., Weuster-Botz, D., and Dietz, H. Biotechnological mass production of DNA origami. *Nature*, 552:84–87, 2017. DOI: 10.1038/nature24650. 24

[66] Ong, L. L., Hanikel, N., Yaghi, O. K., Grun, C., Strauss, M. T., Bron, P., Lai-Kee-Him, J., Schueder, F., Wang, B., Wang, P., et al. Programmable self-assembly of three-dimensional nanostructures from 10,000 unique components. *Nature*, 552:72–77, 2017. DOI: 10.1038/nature24648. 24

[67] Mathur, D. and Medintz, I. L. Analyzing DNA nanotechnology: A call to arms for the analytical chemistry community. *Analytical Chemistry*, 89:2646–2663, 2017. DOI: 10.1021/acs.analchem.6b04033. 25

[68] Shin, J.-S. and Pierce, N. A. A synthetic DNA walker for molecular transport. *Journal of the American Chemical Society*, 126:10834–10835, 2004. DOI: 10.1021/ja047543j. 24, 26

[69] Chen, J., Luo, Z., Sun, C., Huang, Z., Zhou, C., Yin, S., Duan, Y., and Li, Y. Research progress of DNA walker and its recent applications in biosensor. *TrAC Trends in Analytical Chemistry*, 120:115626, 2019. DOI: 10.1016/j.trac.2019.115626. 24

[70] Lee, J. B., Peng, S., Yang, D., Roh, Y. H., Funabashi, H., Park, N., Rice, E. J., Chen, L., Long, R., Wu, M., et al. A mechanical metamaterial made from a DNA hydrogel. *Nature Nanotechnology*, 7:816–820, 2012. DOI: 10.1038/nnano.2012.211. 24, 27

[71] Iinuma, R., Ke, Y., Jungmann, R., Schlichthaerle, T., Woehrstein, J. B., and Yin, P. Polyhedra self-assembled from DNA tripods and characterized with 3D DNA-PAINT. *Science*, 344:65–69, 2014. DOI: 10.1126/science.1250944. 24, 28

[72] Brunsveld, L., Vekemans, J. A. J. M., Hirschberg, J. H. K. K., Sijbesma, R. P., and Meijer, E. W. Hierarchical formation of helical supramolecular polymers via stacking of hydrogen-bonded pairs in water. *Proc. of the National Academy of Sciences*, 99:4977–4982, 2002. DOI: 10.1073/pnas.072659099. 28, 29

[73] Ghadiri, M. R., Granja, J. R., Milligan, R. A., McRee, D. E., and Khazanovich, N. Self-assembling organic nanotubes based on a cyclic peptide architecture. *Nature*, 366:324–327, 1993. DOI: 10.1038/366324a0. 28

[74] Fernandez-Lopez, S., Kim, H.-S., Choi, E. C., Delgado, M., Granja, J. R., Khasanov, A., Kraehenbuehl, K., Long, G., Weinberger, D. A., Wilcoxen, K. M., et al. Antibacterial agents based on the cyclic d,l-α-peptide architecture. *Nature*, 412:452–455, 2001. DOI: 10.1038/35086601. 29

[75] Zhao, Y., Leman, L. J., Search, D. J., Garcia, R. A., Gordon, D. A., Maryanoff, B. E. and Ghadiri, M. R. Self-assembling cyclic d,l-α-peptides as modulators of plasma HDL function. A supramolecular approach toward antiatherosclerotic agents. *ACS Central Science*, 3:639–646, 2017. DOI: 10.1021/acscentsci.7b00154. 29, 30

[76] Yin, L., Agustinus, A. S., Yuvienco, C., Minashima, T., Schnabel, N. L., Kirsch, T., and Montclare, J. K. Engineered coiled-coil protein for delivery of inverse agonist for osteoarthritis. *Biomacromolecules*, 19:1614–1624, 2018. DOI: 10.1021/acs.biomac.8b00158. 29

[77] Katyal, P., Meleties, M., and Montclare, J. K. Self-assembled protein- and peptide-based nanomaterials. *ACS Biomaterials Science and Engineering*, 5:4132–4147, 2019. DOI: 10.1021/acsbiomaterials.9b00408. 29

[78] Gunasekar, S. K., Anjia, L., Matsui, H., and Montclare, J. K. Effects of divalent metals on nanoscopic fiber formation and small molecule recognition of helical proteins. *Advanced Functional Materials*, 22:2154–2159, 2012. DOI: 10.1002/adfm.201101627. 29

[79] McCarthy, H. O., McCaffrey, J., McCrudden, C. M., Zholobenko, A., Ali, A. A., McBride, J. W., Massey, A. S., Pentlavalli, S., Chen, K.-H., Cole, G., et al. Development and characterization of self-assembling nanoparticles using a bio-inspired amphipathic peptide for gene delivery. *Journal of Controlled Release*, 189:141–149, 2014. DOI: 10.1016/j.jconrel.2014.06.048. 30

[80] Bennett, R., Yakkundi, A., McKeen, H. D., McClements, L., McKeogh, T. J., McCrudden, C. M., Arthur, K., Robson, T., and McCarthy, H. O. RALA-mediated delivery of FKBPL nucleic acid therapeutics. *Nanomedicine*, 10:2989–3001, London, 2015. DOI: 10.2217/nnm.15.115. 30

[81] Garmann, R. F., Goldfain, A. M., and Manoharan, V. N. Measurements of the self-assembly kinetics of individual viral capsids around their RNA genome. *Proc. of the National Academy of Sciences*, 116:22485–22490, 2019. DOI: 10.1073/pnas.1909223116. 30

[82] Olson, A. J., Hu, Y. H. E., and Keinan, E. Chemical mimicry of viral capsid self-assembly. *Proc. of the National Academy of Sciences*, 104:20731–20736, 2007. DOI: 10.1073/pnas.0709489104. 30

[83] Butler, P. J. Self-assembly of tobacco mosaic virus: The role of an intermediate aggregate in generating both specificity and speed. *Philosophical Transactions of the Royal Society of London. Series B, Biological Sciences*, 354:537–550, 1999. DOI: 10.1098/rstb.1999.0405. 31

[84] Kegel, W. K. and van der Schoot, P. Physical regulation of the self-assembly of tobacco mosaic virus coat protein. *Biophysical Journal*, 91:1501–1512, 2006. DOI: 10.1529/biophysj.105.072603. 31

[85] Nguyen, H. G., Metavarayuth, K., and Wang, Q. Upregulation of osteogenesis of mesenchymal stem cells with virus-based thin films. *Nanotheranostics*, 2:42–58, 2018. DOI: 10.7150/ntno.19974. 31

[86] Lizotte, P. H., Wen, A. M., Sheen, M. R., Fields, J., Rojanasopondist, P., Steinmetz, N. F., and Fiering, S. In situ vaccination with cowpea mosaic virus nanoparticles suppresses metastatic cancer. *Nature Nanotechnology*, 11:295–303, 2016. DOI: 10.1038/nnano.2015.292. 31, 32

[87] Fuenmayor, J., Gòdia, F., and Cervera, L. Production of virus-like particles for vaccines. *New Biotechnology*, 39:174–180, 2017. DOI: 10.1016/j.nbt.2017.07.010. 31

[88] Peruzzi, P. P. and Chiocca, E. A. A vaccine from plant virus proteins. *Nature Nanotechnology*, 11:214–215, 2016. DOI: 10.1038/nnano.2015.306. 31, 32

[89] Ong, H. K., Tan, W. S., and Ho, K. L. Virus like particles as a platform for cancer vaccine development. *PeerJ*, 5:e4053, 2017. DOI: 10.7717/peerj.4053. 31

[90] Thong, Q. X., Biabanikhankahdani, R., Ho, K. L., Alitheen, N. B., and Tan, W. S. Thermally-responsive virus-like particle for targeted delivery of cancer drug. *Scientific Reports*, 9:3945, 2019. DOI: 10.1038/s41598-019-40388-x. 31, 33

[91] Niu, Z., Bruckman, M. A., Li, S., Lee, L. A., Lee, B., Pingali, S. V., Thiyagarajan, P., and Wang, Q. Assembly of tobacco mosaic virus into fibrous and macroscopic bundled arrays mediated by surface aniline polymerization. *Langmuir*, 23:6719–6724, 2007. DOI: 10.1021/la070096b. 33, 34

[92] Bruckman, M. A., Niu, Z., Li, S., Lee, L. A., Nelson, T. L., Lavigne, J. J., Wang, Q., and Varazo, K. Development of nanobiocomposite fibers by controlled assembly of rod-like tobacco mosaic virus. *NanoBiotechnology*, 3:31–39, 2007. DOI: 10.1007/s12030-007-0004-4. 33

[93] Steinmetz, N. F. and Manchester, M. Playing "nano-lego": VNPs as building blocks for the construction of multi-dimentional arrays. *Viral Nanoparticles*, Jenny Stanford Publishing, New York, 2011. DOI: 10.1201/9780429067457. 33

[94] Rong, J., Lee, L. A., Li, K., Harp, B., Mello, C. M., Niu, Z., and Wang, Q. Oriented cell growth on self-assembled bacteriophage M13 thin films. *Chemical Communications*, pages 5185–5187, 2008. DOI: 10.1039/b811039e. 34, 35

[95] Wargacki, S. P., Pate, B., and Vaia, R. A. Fabrication of 2D ordered films of tobacco mosaic virus (TMV): Processing morphology correlations for convective assembly. *Langmuir*, 24:5439–5444, 2008. DOI: 10.1021/la7040778. 34

[96] What is nanomedicine? https://etp-nanomedicine.eu/about-nanomedicine/what-is-nanomedicine/ 37

[97] National Institutes of Health Office of Strategic Coordination—The Common Fund. Program: Nanomedicine—Overview. https://commonfund.nih.gov/nanomedicine/overview 37

[98] Kim, T. H., Lee, S., and Chen, X. Nanotheranostics for personalized medicine. *Expert Review of Molecular Diagnostics*, 13:257–269, 2013. DOI: 10.1586/erm.13.15. 37

[99] Patra, J. K., Das, G., Fraceto, L. F., Campos, E. V. R., Rodriguez-Torres, M. D. P., Acosta-Torres, L. S., Diaz-Torres, L. A., Grillo, R., Swamy, M. K., Sharma, S., et al. Nano based drug delivery systems: Recent developments and future prospects. *Journal of Nanobiotechnology*, 16:71, 2018. DOI: 10.1186/s12951-018-0392-8. 37

[100] Din, F. U., Aman, W., Ullah, I., Qureshi, O. S., Mustapha, O., Shafique, S., and Zeb, A. Effective use of nanocarriers as drug delivery systems for the treatment of selected tumors. *International Journal of Nanomedicine*, 12:7291–7309, 2017. DOI: 10.2147/ijn.s146315. 37

[101] Nakamura, Y., Mochida, A., Choyke, P. L., and Kobayashi, H. Nanodrug delivery: Is the enhanced permeability and retention effect sufficient for curing cancer? *Bioconjugate Chemistry*, 27:2225–2238, 2016. DOI: 10.1021/acs.bioconjchem.6b00437. 37

[102] Wang, A. Z. EPR or no EPR? The billion-dollar question. *Science Translational Medicine*, 7:294ec112, 2015. DOI: 10.1126/scitranslmed.aac8108. 37

[103] Attia, M. F., Anton, N., Wallyn, J., Omran, Z., and Vandamme, T. F. An overview of active and passive targeting strategies to improve the nanocarriers efficiency to tumour sites. *Journal of Pharmacy and Pharmacology*, 71:1185–1198, 2019. DOI: 10.1111/jphp.13098. 37

[104] Anarjan, F. S. Active targeting drug delivery nanocarriers: Ligands. *Nano-Structures and Nano-Objects*, 19:100370, 2019. 37

[105] Turecek, P. L., Bossard, M. J., Schoetens, F., and Ivens, I. A. PEGylation of biopharmaceuticals: A review of chemistry and nonclinical safety information of approved drugs. *Journal of Pharmaceutical Sciences*, 105:460–475, 2016. DOI: 10.1016/j.xphs.2015.11.015. 37

[106] Poon, W., Zhang, Y.-N., Ouyang, B., Kingston, B. R., Wu, J. L. Y., Wilhelm, S., and Chan, W. C. W. Elimination pathways of nanoparticles. *ACS Nano*, 13:5785–5798, 2019. DOI: 10.1021/acsnano.9b01383. 37

[107] Im, H.-J., England, C. G., Feng, L., Graves, S. A., Hernandez, R., Nickles, R. J., Liu, Z., Lee, D. S., Cho, S. Y., and Cai, W. Accelerated blood clearance phenomenon reduces the passive targeting of PEGylated nanoparticles in peripheral arterial disease. *ACS Applied Materials and Interfaces*, 8:17955–17963, 2016. DOI: 10.1021/acsami.6b05840. 38

[108] Zhou, L.-Q., Li, P., Cui, X.-W., and Dietrich, C. F. Ultrasound nanotheranostics in fighting cancer: Advances and prospects. *Cancer Letters*, 470:204–219, 2020. DOI: 10.1016/j.canlet.2019.11.034. 38

[109] Min, K. H., Min, H. S., Lee, H. J., Park, D. J., Yhee, J. Y., Kim, K., Kwon, I. C., Jeong, S. Y., Silvestre, O. F., Chen, X., et al. pH-Controlled gas-generating mineralized nanoparticles: A theranostic agent for ultrasound imaging and therapy of cancers. *ACS Nano*, 9:134–145, 2015. DOI: 10.1021/nn506210a. 38, 39

[110] Nagamune, T. Biomolecular engineering for nanobio/bionanotechnology. *Nano Convergence*, 4:9, 2017. DOI: 10.1186/s40580-017-0103-4. 38

[111] Xue, Y. Chapter 11—Carbon nanotubes for biomedical applications. *Industrial Applications of Carbon Nanotubes*, Peng, H., Li, Q., and Chen, T., Eds., Elsevier, Boston, 2017. DOI: 10.1016/B978-0-323-41481-4.00011-3. 38

[112] Sireesha, M., Jagadeesh Babu, V., Kranthi Kiran, A. S., and Ramakrishna, S. A review on carbon nanotubes in biosensor devices and their applications in medicine. *Nanocomposites*, 4:36–57, 2018. DOI: 10.1080/20550324.2018.1478765. 38

[113] Çakıroğlu, B. and Özacar, M. A self-powered photoelectrochemical glucose biosensor based on supercapacitor Co3O4-CNT hybrid on TiO2. *Biosensors and Bioelectronics*, 119:34–41, 2018. DOI: 10.1016/j.bios.2018.07.049. 38

[114] Tilmaciu, C.-M. and Morris, M. C. Carbon nanotube biosensors. *Frontiers in Chemistry*, 3:59, 2015. DOI: 10.3389/fchem.2015.00059. 40

[115] Fayazfar, H., Afshar, A., Dolati, M., and Dolati, A. DNA impedance biosensor for detection of cancer, TP53 gene mutation, based on gold nanoparticles/aligned carbon nanotubes modified electrode. *Analytica Chimica Acta*, 836:34–44, 2014. DOI: 10.1016/j.aca.2014.05.029. 40

[116] Langer, R. and Vacanti, J. Tissue engineering. *Science*, 260:920–926, 1993. DOI: 10.1126/science.8493529. 40

[117] Li, J., Esteban-Fernández de Ávila, B., Gao, W., Zhang, L., and Wang, J. Micro/nanorobots for biomedicine: Delivery, surgery, sensing, and detoxification. *Science Robotics*, 2:eaam6431, 2017. DOI: 10.1126/scirobotics.aam6431. 40

[118] Gobaa, S., Gayet, R. V., and Lutolf, M. P. Chapter 3—Artificial niche microarrays for identifying extrinsic cell-fate determinants. *Methods in Cell Biology*, Fletcher, D. A., Doh, J., and Piel, M., Eds., Academic Press, 2018. DOI: 10.1016/bs.mcb.2018.06.012. 40

[119] Charrier, E. E., Pogoda, K., Wells, R. G., and Janmey, P. A. Control of cell morphology and differentiation by substrates with independently tunable elasticity and viscous dissipation. *Nature Communications*, 9:449, 2018. DOI: 10.1038/s41467-018-02906-9. 40

[120] Nguyen, Q. D. and Chung, K.-H. Effect of tip shape on nanomechanical properties measurements using AFM. *Ultramicroscopy*, 202:1–9, 2019. DOI: 10.1016/j.ultramic.2019.03.012. 40

[121] Stevens, M. M. and George, J. H. Exploring and engineering the cell surface interface. *Science*, 310:1135–1138, 2005. DOI: 10.1126/science.1106587. 40

[122] Miller, Y., Ma, B., and Nussinov, R. Polymorphism in self-assembly of peptide-based β-hairpin contributes to network morphology and hydrogel mechanical rigidity. *The Journal of Physical Chemistry B*, 119:482–490, 2015. DOI: 10.1021/jp511485n. 40

[123] Yeo, W.-S., Yousaf, M. N., and Mrksich, M. Dynamic interfaces between cells and surfaces: Electroactive substrates that sequentially release and attach cells. *Journal of the American Chemical Society*, 125:14994–14995, 2003. DOI: 10.1021/ja038265b. 41

[124] Fukuda, J., Khademhosseini, A., Yeh, J., Eng, G., Cheng, J., Farokhzad, O. C., and Langer, R. Micropatterned cell co-cultures using layer-by-layer deposition of extracellular matrix components. *Biomaterials*, 27:1479–1486, 2006. DOI: 10.1016/j.biomaterials.2005.09.015. 41

[125] Gold, L. I. and Pearlstein, E. Fibronectin-collagen binding and requirement during cellular adhesion. *Biochemical Journal*, 186:551–559, 1980. DOI: 10.1042/bj1860551. 41

[126] Rose, S., Prevoteau, A., Elzière, P., Hourdet, D., Marcellan, A., and Leibler, L. Nanoparticle solutions as adhesives for gels and biological tissues. *Nature*, 505:382–385, 2014. DOI: 10.1038/nature12806. 41

[127] Kateb, B. and Heiss, J., Eds., *The Textbook of Nanoneuroscience and Nanoneurosurgery*, CRC Press, Boca Raton, FL, 2014. 41

[128] FAO: The State of Food Security and Nutrition in the World. http://www.fao.org/state-of-food-security-nutrition/en/ 43

[129] USDA: Food Waste FAQs. https://www.usda.gov/foodwaste/faqs 43

[130] FAO: Food Loss and Food Waste. http://www.fao.org/food-loss-and-food-waste/en/ 43

[131] Singh, T., Shukla, S., Kumar, P., Wahla, V., Bajpai, V. K., and Rather, I. A. Application of nanotechnology in food science: Perception and overview. *Frontiers in Microbiology*, 8, 2017. DOI: 10.3389/fmicb.2017.01501. 43

[132] Chaudhry, Q. and Castle, L. Food applications of nanotechnologies: An overview of opportunities and challenges for developing countries. *Trends in Food Science and Technology*, 22:595–603, 2011. DOI: 10.1016/j.tifs.2011.01.001. 43

[133] Joye, I. J., Davidov-Pardo, G., and McClements, D. J. Nanotechnology in food processing. *Encyclopedia of Food and Health*, pages 49–55, Caballero, B., Finglas, P. M., and Toldrá, F., Eds., Academic Press, Oxford, 2016. DOI: 10.1016/B978-0-12-384947-2.00481-5. 43

[134] Zhu, J. and Huang, Q. Chapter 4—Nanoencapsulation of functional food ingredients. *Advances in Food and Nutrition Research*, Lim, L.-T. and Rogers, M., Eds., Academic Press, 2019. 43

[135] Rodrigues, R. M., Ramos, P. E., Cerqueira, M. F., Teixeira, J. A., Vicente, A. A., Pastrana, L. M., Pereira, R. N., and Cerqueira, M. A. Electrosprayed whey protein-based nanocapsules for β-carotene encapsulation. *Food Chemistry*, 314:126157, 2020. DOI: 10.1016/j.foodchem.2019.126157. 43

[136] Karim, M., Fathi, M., and Soleimanian-Zad, S. Nanoencapsulation of cinnamic aldehyde using zein nanofibers by novel needle-less electrospinning: Production, characterization and their application to reduce nitrite in sausages. *Journal of Food Engineering*, 288:110140, 2021. DOI: 10.1016/j.jfoodeng.2020.110140. 43

[137] Akkermans, C., van der Goot, A. J., Venema, P., van der Linden, E., and Boom, R. M. Formation of fibrillar whey protein aggregates: Influence of heat and shear treatment, and resulting rheology. *Food Hydrocolloids*, 22:1315–1325, 2008. DOI: 10.1016/j.foodhyd.2007.07.001. 43

[138] Wang, L., Mu, R.-J., Li, Y., Lin, L., Lin, Z., and Pang, J. Characterization and antibacterial activity evaluation of curcumin loaded konjac glucomannan and zein nanofibril films. *LWT—Food Science and Technology*, 113:108293, 2019. DOI: 10.1016/j.lwt.2019.108293. 43

[139] Loveday, S. M., Wang, X. L., Rao, M. A., Anema, S. G., and Singh, H. β-Lactoglobulin nanofibrils: Effect of temperature on fibril formation kinetics, fibril morphology and the rheological properties of fibril dispersions. *Food Hydrocolloids*, 27:242–249, 2012. DOI: 10.1016/j.foodhyd.2011.07.001. 43

[140] Zhao, H.-S., Ma, Z., and Jing, P. Interaction of soy protein isolate fibrils with betalain from red beetroots: Morphology, spectroscopic characteristics and thermal stability. *Food Research International*, 135:109289, 2020. DOI: 10.1016/j.foodres.2020.109289. 43

[141] Jampílek, J. and Kráľová, K. Nanomaterials applicable in food protection. *Nanotechnology Applications in the Food Industry*, Rai, V. and Bai, J., Eds., CRC Press, Boca Raton, FL, 2018. DOI: 10.1201/9780429488870-5. 44

[142] Nicoletti, M. and Del Serrone, P. Intelligent and smart packaging. *Future Foods*, Mikkola, H., Ed., IntechOpen, 2017. DOI: 10.5772/intechopen.68773. 44

[143] Films and coatings produced from biopolymers and composites. *Biopolymer Engineering in Food Processing*, Telis, V., Ed., CRC Press, Boca Raton, FL, 2012. 44

[144] Ong, Y., Yee, K., Yeang, Q., Zein, S., and Tan, S. Starch based composites for packaging applications. *Handbook of Bioplastics and Biocomposites Engineering Applications*, Kharisov, B., Kharissova O., and Dias, H., Eds., John Wiley & Sons, Inc., 2014. DOI: 10.1002/9781118203699.ch8. 44

[145] Lee, S. Y., Lee, S. J., Choi, D. S., and Hur, S. J. Current topics in active and intelligent food packaging for preservation of fresh foods. *Journal of the Science of Food and Agriculture*, 95:2799–2810, 2015. DOI: 10.1002/jsfa.7218. 44

[146] Mohammadi, M., Mirabzadeh, S., Shahvalizadeh, R., and Hamishehkar, H. Development of novel active packaging films based on whey protein isolate incorporated with chitosan nanofiber and nano-formulated cinnamon oil. *International Journal of Biological Macromolecules*, 149:11–20, 2020. DOI: 10.1016/j.ijbiomac.2020.01.083. 44

[147] Sun, J., Jiang, H., Wu, H., Tong, C., Pang, J., and Wu, C. Multifunctional bionanocomposite films based on konjac glucomannan/chitosan with nano-ZnO and mulberry anthocyanin extract for active food packaging. *Food Hydrocolloids*, 107:105942, 2020. DOI: 10.1016/j.foodhyd.2020.105942. 44

[148] Ramos, M., Jiménez, A., Peltzer, M., and Garrigós, M. C. Development of novel nanobiocomposite antioxidant films based on poly (lactic acid) and thymol for active packaging. *Food Chemistry*, 162:149–155, 2014. DOI: 10.1016/j.foodchem.2014.04.026. 44

[149] Bi, F., Zhang, X., Liu, J., Yong, H., Gao, L., and Liu, J. Development of antioxidant and antimicrobial packaging films based on chitosan, D-α-tocopheryl polyethylene glycol 1000 succinate and silicon dioxide nanoparticles. *Food Packaging and Shelf Life*, 24:100503, 2020. DOI: 10.1016/j.fpsl.2020.100503. 44

[150] Kalpana, S., Priyadarshini, S. R., Maria Leena, M., Moses, J. A., and Anandharamakrishnan, C. Intelligent packaging: Trends and applications in food systems. *Trends in Food Science and Technology*, 93:145–157, 2019. DOI: 10.1016/j.tifs.2019.09.008. 44

[151] Zhang, C., Yin, A.-X., Jiang, R., Rong, J., Dong, L., Zhao, T., Sun, L.-D., Wang, J., Chen, X., and Yan, C.-H. Time—Temperature indicator for perishable products based on kinetically programmable Ag overgrowth on Au nanorods. *ACS Nano*, 7:4561–4568, 2013. DOI: 10.1021/nn401266u. 44, 46

[152] About FoodSafety.gov. https://www.foodsafety.gov/about 44

[153] BE Disclosure. https://www.ams.usda.gov/rules-regulations/be 44

[154] Kumar, V., Guleria, P., and Mehta, S. K. Nanosensors for food quality and safety assessment. *Environmental Chemistry Letters*, 15:165–177, 2017. DOI: 10.1007/s10311-017-0616-4. 45

[155] Sarghini, F. and Marra, F. Nanosensors in the food industry. *Nanotechnology Applications in the Food Industry*, Rai, V. and Bai, J., Eds., CRC Press, Boca Raton, FL, 2018. DOI: 10.1201/9780429488870. 45

[156] Thevendran, R., Sarah, S., Tang, T.-H., and Citartan, M. Strategies to bioengineer aptamer-driven nanovehicles as exceptional molecular tools for targeted therapeutics: A review. *Journal of Controlled Release*, 323:530–548, 2020. DOI: 10.1016/j.jconrel.2020.04.051. 45

[157] Verdian, A. Apta-nanosensors for detection and quantitative determination of acetamiprid—A pesticide residue in food and environment. *Talanta*, 176:456–464, 2018. DOI: 10.1016/j.talanta.2017.08.070. 45

[158] Jia, Y., Wu, F., Liu, P., Zhou, G., Yu, B., Lou, X., and Xia, F. A label-free fluorescent aptasensor for the detection of Aflatoxin B1 in food samples using AIEgens and graphene oxide. *Talanta*, 198:71–77, 2019. DOI: 10.1016/j.talanta.2019.01.078. 45, 47

[159] Pathakoti, K., Manubolu, M., and Hwang, H.-M. Chapter 48—Nanotechnology applications for environmental industry. *Handbook of Nanomaterials for Industrial Applications*, Mustansar Hussain, C., Ed., Elsevier, 2018. DOI: 10.1016/B978-0-12-813351-4.00050-X. 49

[160] Burger, J., Gochfeld, M., Kosson, D. S., Brown, K. G., Salisbury, J. A., and Jeitner, C. Risk to ecological resources following remediation can be due mainly to increased resource value of successful restoration: A case study from the Department of Energy's Hanford Site. *Environmental Research*, 186:109536, 2020. DOI: 10.1016/j.envres.2020.109536. 49

[161] Qu, X., Alvarez, P. J. J., and Li, Q. Applications of nanotechnology in water and wastewater treatment. *Water Research*, 47:3931–3946, 2013. DOI: 10.1016/j.watres.2012.09.058. 49

[162] Kusworo, T. D., Ariyanti, N., and Utomo, D. P. Effect of nano-TiO2 loading in polysulfone membranes on the removal of pollutant following natural-rubber wastewater treatment. *Journal of Water Process Engineering*, 35:101190, 2020. DOI: 10.1016/j.jwpe.2020.101190. 49

[163] Grespania launch H&C TILES© with HYDROTECT©. https://gb.toto.com/press/press/?tx_ttnews%5Btt_news%5D=424&cHash=9f7f318b35ea99f96f9f6276d1e0f90b 49

[164] Pawar, M., Topcu Sendoğdular, S., and Gouma, P. A brief overview of TiO_2 photocatalyst for organic dye remediation: Case study of reaction mechanisms involved in Ce-TiO_2 photocatalysts system. *Journal of Nanomaterials*, 5953609, 2018. DOI: 10.1155/2018/5953609. 50

[165] U.S. EPA: Ecological Risk Assessment Glossary of Terms. https://ofmpub.epa.gov/sor_internet/registry/termreg/searchandretrieve/glossariesandkeywordlists/search.do?details=&glossaryName=Eco%20Risk%20Assessment%20Glossary 49

[166] Kharisov, B. I., Dias, H. V. R., and Kharissova, O. V. Nanotechnology-based remediation of petroleum impurities from water. *Journal of Petroleum Science and Engineering*, 122:705–718, 2014. DOI: 10.1016/j.petrol.2014.09.013. 49

[167] Karaaslan, M. A., Kadla, J. F., and Ko, F. K. Chapter 5—Lignin-based aerogels. *Lignin in Polymer Composites*, Faruk, O. and Sain, M., Eds., William Andrew Publishing, 2016. DOI: 10.1016/B978-0-323-35565-0.00005-9. 49

[168] Thai, Q. B., Nguyen, S. T., Ho, D. K., Tran, T. D., Huynh, D. M., Do, N. H. N., Luu, T. P., Le, P. K., Le, D. K., Phan-Thien, N., et al. Cellulose-based aerogels from sugarcane bagasse for oil spill-cleaning and heat insulation applications. *Carbohydrate Polymers*, 228:115365, 2020. DOI: 10.1016/j.carbpol.2019.115365. 49, 50

[169] Jiang, D., Zeng, G., Huang, D., Chen, M., Zhang, C., Huang, C., and Wan, J. Remediation of contaminated soils by enhanced nanoscale zero valent iron. *Environmental Research*, 163:217–227, 2018. DOI: 10.1016/j.envres.2018.01.030. 49

[170] Yee, K., Yeang, Q., Ong, Y., Vadivelu, V., and Tan, S. Water remediation using nanoparticle and nanocomposite membranes. *Nanomaterials for Environmental Protection*, Kharisov, B., Kharissova O., and Dias, H., Eds., John Wiley & Sons, Inc., 2014. DOI: 10.1002/9781118845530.ch17. 51

[171] Clough, S. The potential ecological hazard of nanomaterials. *Nanotechnology and the Environment*, CRC Press, Boca Raton, FL, 2009. 51

[172] Stander, L. and Theodore, L. Environmental implications of nanotechnology—An update. *International Journal of Environmental Research and Public Health*, 8:470–479, 2011. DOI: 10.3390/ijerph8020470. 51

[173] Inshakova, E. and Inshakov, O. World market for nanomaterials: Structure and trends. *MATEC Web Conferences*, 129:02013, 2017. DOI: 10.1051/matecconf/201712902013. 51

[174] Abbas, Q., Yousaf, B., Amina, Ali, M. U., Munir, M. A. M., El-Naggar, A., Rinklebe, J., and Naushad, M. Transformation pathways and fate of engineered nanoparticles (ENPs) in distinct interactive environmental compartments: A review. *Environment International*, 138:105646, 2020. DOI: 10.1016/j.envint.2020.105646. 51

[175] Zeng, J., Xu, P., Chen, G., Zeng, G., Chen, A., Hu, L., Huang, Z., He, K., Guo, Z., Liu, W., et al. Effects of silver nanoparticles with different dosing regimens and exposure media on artificial ecosystem. *Journal of Environmental Sciences*, 75:181–192, 2019. DOI: 10.1016/j.jes.2018.03.019. 51

[176] Wang, J., Koo, Y., Alexander, A., Yang, Y., Westerhof, S., Zhang, Q., Schnoor, J. L., Colvin, V. L., Braam, J., and Alvarez, P. J. J. Phytostimulation of poplars and arabidopsis exposed to silver nanoparticles and Ag+ at sublethal concentrations. *Environmental Science and Technology*, 47:5442–5449, 2013. DOI: 10.1021/es4004334. 52

[177] Ferdous, Z. and Nemmar, A. Health impact of silver nanoparticles: A review of the biodistribution and toxicity following various routes of exposure. *International Journal of Molecular Sciences*, 21:2375, 2020. DOI: 10.3390/ijms21072375. 52

[178] Romero-Franco, M., Godwin, H. A., Bilal, M., and Cohen, Y. Needs and challenges for assessing the environmental impacts of engineered nanomaterials (ENMs). *Beilstein Journal of Nanotechnology*, 8:989–1014, 2017. DOI: 10.3762/bjnano.8.101. 52, 53

[179] ISO 14040:2006. https://www.iso.org/standard/37456.html 53

[180] Shatkin, J. Nano LCRA: An adaptive screening-level life cycle risk-assessment framework for nanotechnology. *Nanotechnology: Health and Environmental Risks*, 2nd ed., CRC Press, Boca Raton, FL, 2012. 53

[181] Shatkin, J. A. and Kim, B. Cellulose nanomaterials: Life cycle risk assessment, and environmental health and safety roadmap. *Environmental Science: Nano*, 2:477–499, 2015. DOI: 10.1039/c5en00059a. 53

[182] Rauscher, H., Rasmussen, K., and Sokull-Klüttgen, B. Regulatory aspects of nanomaterials in the EU. *Chemie Ingenieur Technik*, 89:224–231, 2017. DOI: 10.1002/cite.201600076.

[183] U.S. CDC NIOSH: Nanotechnology—Guidance and Publications. https://www.cdc.gov/niosh/topics/nanotech/pubs.html

[184] USDA National Organic Program: Policy Memorandum. https://www.ams.usda.gov/sites/default/files/media/NOP-PM-15--2-Nanotechnology.pdf

[185] U.S. EPA: Control of Nanoscale Materials under the Toxic Substances Control Act. https://www.epa.gov/reviewing-new-chemicals-under-toxic-substances-control-act-tsca/control-nanoscale-materials-under

[186] U.S. EPA: Technical Fact Sheet—Nanomaterials. https://www.epa.gov/sites/production/files/2014--03/documents/ffrrofactsheet_emergingcontaminant_nanomaterials_jan2014_final.pdf

[187] U.S. EPA: Nanotechnology White Paper. https://www.epa.gov/sites/production/files/2015--01/documents/nanotechnology_whitepaper.pdf

[188] U.S. FDA: Nanotechnology Guidance Documents. https://www.fda.gov/science-research/nanotechnology-programs-fda/nanotechnology-guidance-documents

[189] National Nanotechnology Initiative: Environmental, Health, and Safety (EHS) Research Strategy. https://www.nano.gov/sites/default/files/pub_resource/nni_2011_ehs_research_strategy.pdf

Author's Biography

WUJIE ZHANG

Wujie Zhang, Ph.D., is an associate professor in the Biomolecular Engineering Program at the Milwaukee School of Engineering (MSOE). He received a Ph.D. degree in biomedical engineering from the University of South Carolina and M.S. and B.S. degrees in food science and engineering from the University of Shanghai for Science and Technology. Dr. Zhang's scholarly work and research span biomaterials, micro/nanotechnology, cellular and tissue engineering, drug delivery, and cancer treatment. He has been teaching the Molecular Nanotechnology course for the past eight years. His current nanotechnology research projects include biosynthesis of metallic nanoparticles, electrospun nanofibers, and nanomedicine. Dr. Zhang has received numerous teaching and research awards, including ASEE Prism magazine's "20 under 40" (2018).

Printed in the United States
by Baker & Taylor Publisher Services